# Interactive Reader and Study Guide

## Holt Social Studies

# Western World

**HOLT, RINEHART AND WINSTON**

A Harcourt Education Company

Orlando • **Austin** • New York • San Diego • London

If you have received these materials as examination copies free of charge, Holt, Rineha<br>and Winston retains title to the materials and they may not be resold.  Resale of<br>examination copies is strictly prohibited.

Possession of this publication in print format does not entitle users to convert this<br>publication, or any portion of it, into electronic format.

ISBN 0-03-078711-4

3 4 5 6 7 8 9  0956  12 11 10 09

# Contents





     

# How to Use this Book

The *Interactive Reader and Study Guide* was developed to help you get the most from your world geography course. Using this book will help you master the content of the course while developing your reading and vocabulary skills. Reviewing the next few pages before getting started will make you aware of the many useful features of this book.

*Chapter Summary pages help you connect with the big picture. Studying them will keep you focused on the information you will need to be successful on your exams.*

**Graphic organizers help you to summarize each chapter.**
- **Some have blanks you will need to fill.**
- **Others have been completed for you.**

**Either way, they are a valuable study tool to help you prepare for important tests.**

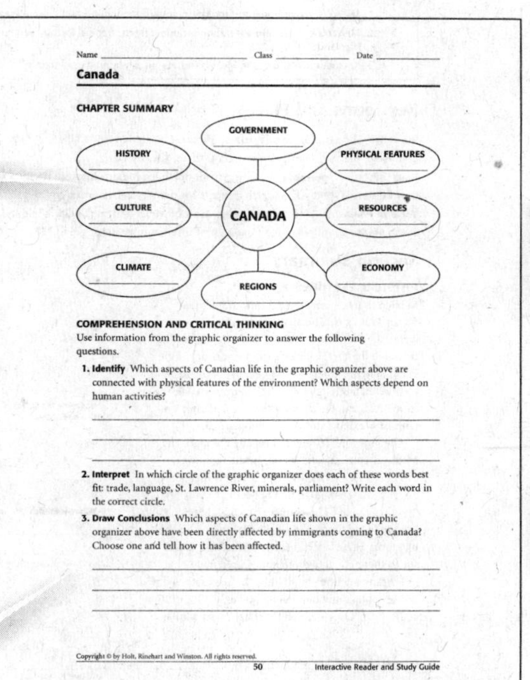


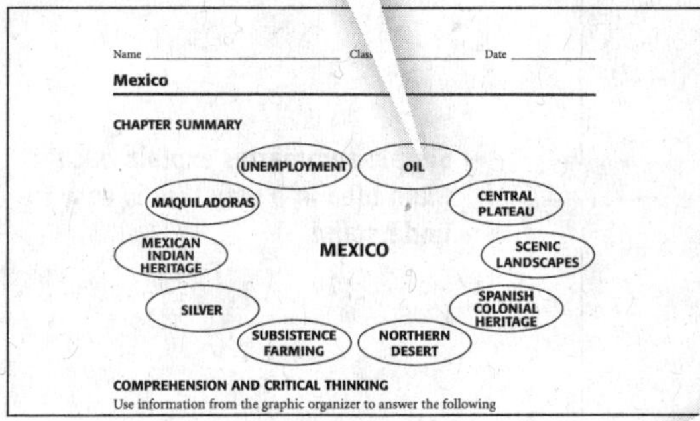


**Answering each question will help you to understand the graphic organizer and ensure that you fully comprehend the content from the chapter.**

*Section Summary pages allow you to interact easily with the content and Key Terms from each section.*

**Main Ideas statements from your textbook focus your attention as you read the summaries.**

**Clearly labeled page headers make navigating the book extraordinarily simple.**

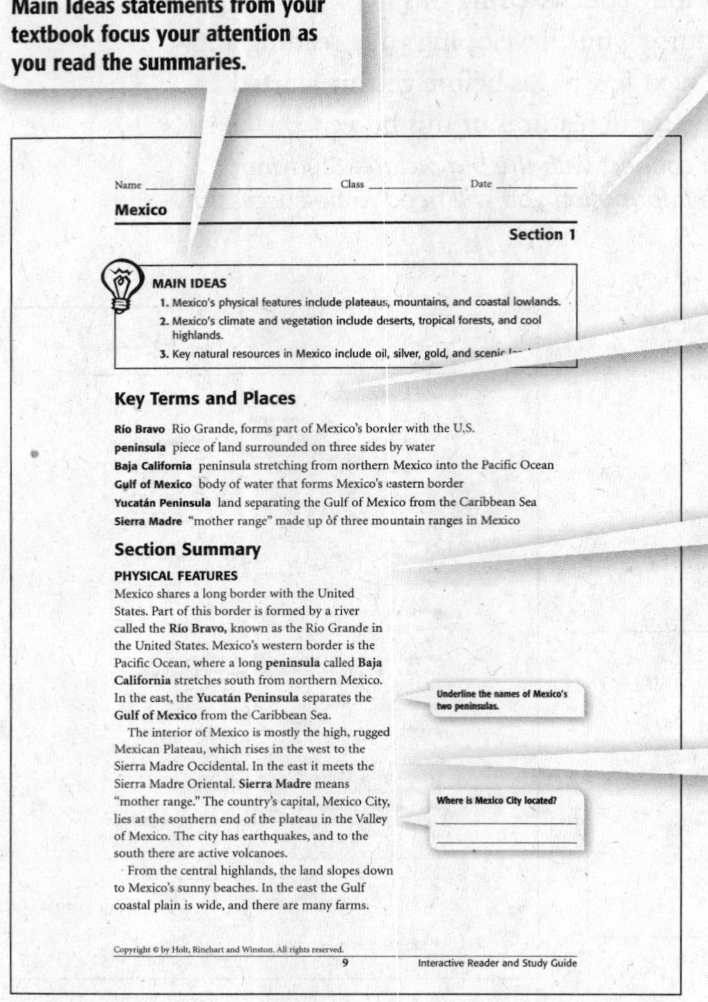


**The Key Terms from your textbook are provided with their definitions, making studying them easier.**

**Headings under each section summary relate to each heading in the textbook, making it easy for you to find the material you need.**

**Simple summaries explain each Main Idea in a way that is easy to understand.**

*Notes throughout the margins help you to interact with the content and understand the information you are reading.*

*Challenge Activities following each section will bring the material to life and further develop your analysis skills.*

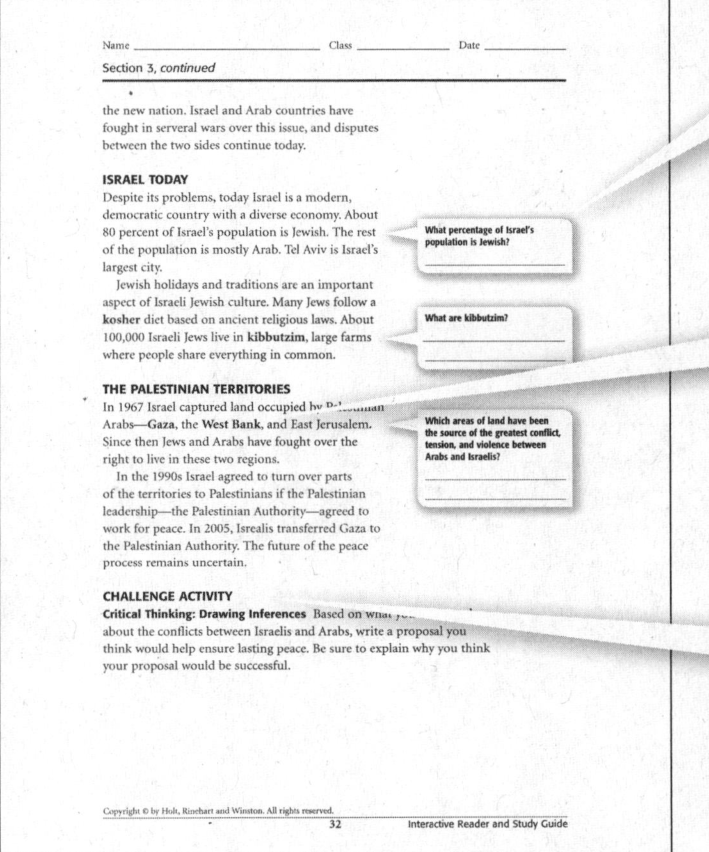


**Be sure to read all notes and answer all of the questions in the margins. They have been written specifically to help you keep track of important information. Your answers to these questions will help you to study for your tests.**

**The Key Terms from your textbook have been boldfaced, allowing you to find and study them quickly.**

**Challenge Activities ask you to think critically about the material from the section. Completing each of these exercises will increase your understanding of the material and develop your analysis skills.**

# A Geographer's World

## CHAPTER SUMMARY

<table>
<tr><td colspan="3" align="center">GEOGRAPHY</td></tr>
<tr><td>Looks at the world in new ways...</td><td>Uses two systems to organize...</td><td>Has two main branches and other branches...</td></tr>
<tr><td>Definition<br>_______________________<br>_______________________<br>_______________________</td><td>Five Themes<br>—location, place, regions, human environment, interaction, movement</td><td>Physical<br>—study of world's physical features</td></tr>
<tr><td>Tools<br>—Maps, globe, other tools</td><td>Six Elements<br>—spatial, places and regions, physical, human, environment and society, uses of geography</td><td>Human<br>—study of world's people and cultures</td></tr>
<tr><td>Levels<br>—local, regional, global</td><td></td><td>Other Branches<br>—cartography, hydrology, meteorology, others</td></tr>
</table>

## COMPREHENSION AND CRITICAL THINKING

Use information from the graphic organizer to answer the following questions.

**1. Identify** Read the three definitions below. Write the one that best defines geography on the lines under the first column. Geography is the study of...

    regions and their physical features.

    the world, people, and their landscapes.

    how different human cultures developed.

**2. Analyze** Give an example of how the five themes and six elements are similar.

_________________________________________________________________

_________________________________________________________________

**3. Interpret** If a geographer were making a map to show the size and height of a hilly area, what branches of geography would this involve?

_________________________________________________________________

**4. Draw a Conclusion** How does geography help us to better understand the world and its peoples?

_________________________________________________________________

# A Geographer's World

**Section 1**

**MAIN IDEAS**

1. Geography is the study of the world, its people, and the landscapes they create.
2. Geographers look at the world in many different ways.
3. Maps and other tools help geographers study the planet.

## Key Terms and Places

**geography**  the study of the world, its people, and the landscapes they create

**landscape**  the human and physical features that make a place unique

**social science**  a field that studies people and the relationships among them

**region**  a part of the world with one or more common features distinguishing it from surrounding areas

**map**  a flat drawing that shows part of Earth's surface

**globe**  a spherical model of the entire planet

## Section Summary

### WHAT IS GEOGRAPHY?

For every place on Earth, you can ask questions to learn about it: What does the land look like? What is the weather like? What are people's lives like? Asking questions like these is how you study geography. **Geography** is the study of the world, its people, and the **landscapes** they create.

Geographers (people who study geography) ask questions about how the world works. For example, they may ask why a place gets tornadoes. To find answers, they gather data by observing and measuring. In this way, geography is like science.

Geography can also be like a social science. **Social science** studies people and how they relate to each other. This information cannot be measured in the same way. To study people, geographers may visit places and talk to the people about their lives.

> Underline the sentence that states how geography is like science.





Interactive Reader and Study Guide

**Section 1, *continued***

## LOOKING AT THE WORLD

Geographers must look carefully at the world around them. Depending on what they want to learn, they look at the world at different levels.

Geographers may study at the local level, such as a city or town. They may ask why people live there, what work they do, and how they travel. They can help a town or city plan improvements.

Geographers may also study at the regional level. A **region** is an area with common features. A region may be big or small. Its features make it different from areas around it. The features may be physical (such as mountains) or human (such as language).

Sometimes geographers study at the global level. They study how people interact all over the world. Geographers can help us learn how people's actions affect other people and places. For example, they may ask how one region influences other regions.

> Circle the three levels that geographers study.

## THE GEOGRAPHER'S TOOLS

Geographers need tools to do their work. Often, they use maps and globes. A **map** is a flat drawing that shows Earth's surface. A **globe** is a spherical (round) model of the whole planet.

Maps and globes both show what Earth looks like. Because a globe is round, it can show Earth as it really is. To show the round Earth on a flat map, some details have to change. For example, a place's shape may change a little. But maps have benefits. They are easier to work with. They can also show small areas, such as cities, better.

Geographers also use other tools, such as satellite images, computers, notebooks, and tape recorders.

> In what way are maps and globes similar?
>
> ________________________________
>
> ________________________________

> Underline two sentences that tell the benefits of using maps.

## CHALLENGE ACTIVITY

**Critical Thinking: Solving Problems** Pick a foreign country you would like to study. You want to develop the most complete picture possible of this place and its people. Make a list of questions to ask and tools you would use to find the answers.



 Interactive Reader and Study Guide

# A Geographer's World

**Section 2**

> **MAIN IDEAS**
> 1. The five themes of geography help us organize our studies of the world.
> 2. The six essential elements of geography highlight some of the subject's most important ideas.

## Key Terms and Places

**absolute location**  a specific description of where a place is

**relative location**  a general description of where a place is

**environment**  an area's land, water, climate, plants and animals, and other physical features

## Section Summary

### THE FIVE THEMES OF GEOGRAPHY

Geographers use themes in their work. A theme is a topic that is common throughout a discussion or event. Many holidays have a theme, such as the flag and patriotism on the Fourth of July.

There are five major themes of geography: Location, Place, Human-Environment Interaction, Movement, and Regions. Geographers can use these themes in almost everything they study.

Location describes where a place is. This may be specific, such as an address. This is called an **absolute location**. It may also be general, such as saying the United States is north of Central America. This is called a **relative location**.

Place refers to an area's landscape. The landscape is made up of the physical and human features of a place. Together, these features give a place its own identity apart from other places.

Human-Environment Interaction studies how people and their environment affect each other. The **environment** includes an area's physical features, such as land, water, weather, and animals.

> List the five major themes of geography:
>
> _______________________
>
> _______________________
>
> _______________________
>
> _______________________
>
> _______________________



   Interactive Reader and Study Guide

**Section 2, *continued***

Geographers study how people change their environment (by building, for example). They also study how the environment causes people to adapt (by dressing for the weather, for example).

Movement involves learning about why and how people move. Do they move for work or pleasure? Do they travel by roads or other routes?

Studying Regions helps geographers learn how places are alike and different. This also helps them learn why places developed the way they did.

> **Describe two ways that people and their environment affect each other.**
> ___________________________
> ___________________________
> ___________________________
> ___________________________
> ___________________________

## THE SIX ESSENTIAL ELEMENTS

It is important to organize how you study geography, so you get the most complete picture of a place. Using the five major themes can help you do this. Using the six essential elements can, also.

> **What do the five themes and six elements of geography help you do? Underline the sentence that explains this.**

Geographers and teachers created the six elements from eighteen basic ideas, called standards. The standards say what everyone should understand about geography. Each element groups together the standards that are related to each other.

The six elements are: The World in Spatial Terms (spatial refers to where places are located); Places and Regions; Physical Systems; Human Systems; Environment and Society; Uses of Geography. The six elements build on the five themes, so some elements and themes are similar. Uses of Geography is not part of the five themes. It focuses on how people can use geography to learn about the past and present, and plan for the future.

## CHALLENGE ACTIVITY

**Critical Thinking: Analyze** Analyze a place you regularly visit, such as a vacation spot or a park in your neighborhood. Write a question about the place for each geography theme to help someone not familiar with the themes understand them.

# A Geographer's World

## Section 3

**MAIN IDEAS**

1. Physical geography is the study of landforms, water bodies, and other physical features.
2. Human geography focuses on people, their cultures, and the landscapes they create.
3. Other branches of geography examine specific aspects of the physical or human world.

## Key Terms and Places

**physical geography**  the study of the world's physical features, such as landforms, bodies of water, climates, soils, and plants

**human geography**  the study of the world's people, communities, and landscapes

**cartography**  the science of making maps

**meteorology**  the study of weather and what causes it

## Section Summary

### PHYSICAL GEOGRAPHY

The field of geography has many branches, or divisions. Each branch has a certain focus. No branch alone gives us a picture of the whole world. When looked at together, the different branches help us understand Earth and its people better.

Geography has two main branches: physical geography and human geography. **Physical geography** is the study of the world's physical features, such as landforms, bodies of water, and weather.

Physical geographers ask questions about Earth's many physical features: Where are the mountains and flat areas? Why are some areas rainy and others dry? Why do rivers flow a certain way? To get their answers, physical geographers measure features— such as heights of mountains and temperatures of places.

> What do the different branches of geography help us do when they are looked at together? Underline the sentence that answers this.

> List the two main branches of geography:
>
> _______________________
>
> _______________________

**Section 3,** *continued*

---

Physical geography has important uses. It helps us understand how the world works. It also helps us predict and prepare for dangers, such as big storms.

## HUMAN GEOGRAPHY

**Human geography** is the study of the world's people, communities, and landscapes. It is the other main branch of geography.

Human geographers study people in the past or present. They ask why more people live in some places than in others. They also ask other questions, such as what kinds of work people do.

People all over the world are very different, so human geographers often study a smaller topic. They might study people in one region, such as central Africa. They might study one part of people's lives in different regions, such as city life.

Human geography also has important uses. It helps us learn how people meet basic needs for food, water, and shelter. It helps people improve their lives. It can also help protect the environment.





## OTHER FIELDS OF GEOGRAPHY

Other branches of geography study one aspect of the world. Some of these are smaller parts of physical geography or of human geography.

Here are a few other branches to know about. **Cartography** is the science of making maps. Hydrology is the study of water on Earth. **Meteorology** is the study of weather and what causes it.



## CHALLENGE ACTIVITY

**Critical Thinking: Drawing Inferences** Examine a map of an unfamiliar city using a road atlas or an online map. Write a paragraph telling a visitor what physical and human features to look for in each quadrant (NE, SE, NW, SW).





Interactive Reader and Study Guide

# Planet Earth

## CHAPTER SUMMARY

| CAUSE | | EFFECT |
|---|---|---|
| Earth's rotation | → | day and night |
| Earth's revolution | → | the seasons |
| the water cycle | → | a constant supply of water on Earth |
| _________________________ | → | water shortages |
| continental drift | → | creates landforms |
| wind, water, and ice | → | _________________________ |

## COMPREHENSION AND CRITICAL THINKING

Use information from the graphic organizer to answer the following questions.

**1. Recall** What is the effect of Earth's movement in space?

_______________________________________________________________

_______________________________________________________________

**2. Interpret** Which two of the following words belong in the graphic organizer above: drought, tilt, precipitation, erosion? Write the two words in the spaces provided.

**3. Synthesis** Under what general theme do all of these cause-and-effect relationships fall?

_______________________________________________________________

_______________________________________________________________

# Planet Earth

## Section 1

**MAIN IDEAS**

1. Earth's movement affects the amount of energy we receive from the sun.
2. Earth's seasons are caused by the planet's tilt.

## Key Terms

**solar energy** energy from the sun

**rotation** one complete spin of Earth on its axis

**revolution** one trip of Earth around the sun

**latitude** the distance north or south of Earth's equator

**tropics** regions close to the equator

## Section Summary

### EARTH'S MOVEMENT

Energy from the sun, or **solar energy**, is necessary for life on Earth. It helps plants grow and provides light and heat. Several factors affect the amount of solar energy Earth receives. These are rotation, revolution, tilt, and latitude.

> Name the four factors that affect the amount of solar energy Earth receives.
> _______________________________
> _______________________________

Earth's axis is an imaginary rod running from the North Pole to the South Pole. Earth spins around on its axis. One complete **rotation** takes 24 hours, or one day. Solar energy reaches only half of the planet at a time. As Earth rotates, levels of solar energy change. The half that faces the sun receives light and heat and is warmer. The half that faces away from the sun is darker and cooler.

> What would happen if Earth did not rotate?
> _______________________________
> _______________________________

As Earth rotates, it also moves around the sun. Earth completes one **revolution** around the sun every year, in 365 1/4 days. Every four years an extra day is added to February. This makes up for the extra quarter of a day.

> Underline the sentence that describes Earth's revolution around the sun.

Earth's axis is tilted, not straight up and down. At different times of year, some locations tilt toward the sun. They get more solar energy than locations tilted away from the sun.

> Where you live, does more solar energy reach Earth in winter or in summer?
> _______________________________



        Interactive Reader and Study Guide

**Section 1, continued**

---

**Latitude** refers to imaginary lines that run east and west around the planet, north and south of the Earth's equator. Areas near the equator receive direct rays from the sun all year and have warm temperatures. Higher latitudes receive fewer direct rays and are cooler.

> Why are areas near the equator warmer than those in higher latitudes?
> ___________________________

## THE SEASONS

Many locations on Earth have four seasons: winter, spring, summer, and fall. These are based on temperature and how long the days are.

The seasons change because of the tilt of Earth's axis. In summer the Northern Hemisphere is tilted toward the sun. It receives more solar energy than during the winter, when it is tilted away from the sun.

Because Earth's axis is tilted, the hemispheres have opposite seasons. Winter in the Northern Hemisphere is summer in the Southern Hemisphere. During the fall and spring, the poles point neither toward nor away from the sun. In spring, temperatures rise and days become longer as summer approaches. In fall the opposite occurs.

> What would the seasons be like in the Northern and Southern hemispheres if Earth's axis weren't tilted?
> ___________________________

In some regions, the seasons are tied to rainfall instead of temperature. One of these regions, close to the equator, is the **tropics**. There, winds bring heavy rains from June to October. The weather turns dry in the tropics from November to January.

> Circle the name of the warm region near the equator.

## CHALLENGE ACTIVITY

**Critical Thinking: Drawing Conclusions** Imagine that you are a travel agent. One of your clients is planning a trip to Argentina in June, and another is planning a trip to Chicago in August. What kinds of clothing would you suggest they pack for their trips and why?

## Planet Earth

Section 2

**MAIN IDEAS**

1. Salt water and freshwater make up Earth's water supply.
2. In the water cycle, water circulates from Earth's surface to the atmosphere and back again.
3. Water plays an important role in people's lives.

## Key Terms

**freshwater** water without salt

**glacier** large area of slow-moving ice

**surface water** water that is stored in Earth's streams, rivers, and lakes

**precipitation** water that falls to Earth's surface as rain, snow, sleet, or hail

**groundwater** water found below Earth's surface

**water vapor** water that occurs in the air as an invisible gas

**water cycle** the circulation of water from Earth's surface to the atmosphere and back

**drought** a long period of lower-than-normal precipitation

## Section Summary

### EARTH'S WATER SUPPLY

Approximately three-quarters of Earth's surface is covered with water. There are two kinds of water—salt water and **freshwater**. About 97 percent of Earth's water is salt water. Most of it is in the oceans, seas, gulfs, bays, and straits. Some lakes, such as the Great Salt Lake in Utah, also contain salt water.

> Circle the places where we find salt water.

Salt water cannot be used for drinking. Only freshwater is safe to drink. Freshwater is found in lakes and rivers and stored underground. Much is frozen in **glaciers**. Freshwater is also found in the ice of the Arctic and Antarctic regions.

> Use an atlas or a globe to locate the Great Salt Lake, the Arctic, and the Antarctic regions.

One form of freshwater is **surface water**. This is stored in streams, lakes, and rivers. Streams form when **precipitation** falls to Earth as rain, snow, sleet, or hail. These streams then flow into larger streams and rivers.

Most freshwater is stored underground. **Groundwater** bubbles to the surface in springs or can be reached by digging deep holes, or wells.

## THE WATER CYCLE

Water can take the form of a liquid, gas, or solid. In its solid form, water is snow and ice. Liquid water is rain or water found in lakes and rivers. **Water vapor** is an invisible form of water in the air.

Water is always moving. When water on Earth's surface heats up, it evaporates and turns into water vapor. It then rises from Earth into the atmosphere. When it cools down, it changes from water vapor to liquid. Droplets of water form clouds. When they get heavier, these droplets fall to Earth as precipitation. This process of evaporation and precipitation is called the **water cycle**.

Some precipitation is absorbed into the soil as groundwater. The rest flows into streams, rivers, and oceans.

## WATER AND PEOPLE

Problems with water include shortages, pollution, and flooding. Shortages are caused by overuse and by **drought**, when there is little or no precipitation for a long time. Chemicals and waste can pollute water. Heavy rains can cause flooding.

Water quenches our thirst and allows us to have food to eat. It is an important source of energy. Water also provides recreation, making our lives richer and more enjoyable. Water is essential for life on Earth.

## CHALLENGE ACTIVITY

**Critical Thinking: Solving Problems** You are campaigning for public office. Write a speech describing three actions you plan to take to protect supplies of freshwater.

> Underline the words that define water vapor.

> What are the two main processes of the water cycle?
>
> _______________________
>
> _______________________

> What water problems affect human beings?
>
> _______________________
>
> _______________________

# Planet Earth

## Section 3

### MAIN IDEAS

1. Earth's surface is covered by many different landforms.
2. Forces below Earth's surface build up our landforms.
3. Forces on the planet's surface shape Earth's landforms.
4. Landforms influence people's lives and culture.

## Key Terms

**landforms**  shapes on Earth's surface, such as hills or mountains

**continents**  large landmasses

**plate tectonics**  a theory suggesting that Earth's surface is divided into more than 12 slow-moving plates, or pieces of Earth's crust

**lava**  magma, or liquid rock, that reaches Earth's surface

**earthquake**  sudden, violent movement of Earth's crust

**weathering**  the process of breaking rock into smaller pieces

**erosion**  the movement of sediment from one location to another

## Section Summary

### LANDFORMS

Geographers study **landforms** such as mountains, valleys, plains, islands, and peninsulas. They study how landforms are made and how they influence people.

> **Give two examples of landforms.**
> ___________________________
> ___________________________

### FORCES BELOW EARTH'S SURFACE

Below Earth's surface, or crust, is a layer of liquid and a solid core. The planet has seven **continents**, large landmasses made of Earth's crust. All of Earth's crust rests on 12 plates. These plates are constantly in motion. Geographers call the study of these moving pieces of crust **plate tectonics**.

All of these plates move at different speeds and in different directions. As they move, they shape Earth's landforms. Plates move in three ways: They collide, they separate, and they slide past each other.

The energy of colliding plates creates new landforms. When two ocean plates collide, they may

> **Look in an atlas. How might the theory of plate tectonics explain the shapes of North America, South America, and Africa?**
> ___________________________



     Interactive Reader and Study Guide

form deep valleys on the ocean's floor. When ocean plates collide with continental plates, mountain ranges are formed. Mountains are also created when two continental plates collide.

When plates separate, usually on the ocean floor, they cause gaps in the planet's crust. Magma, or liquid rock, rises through the cracks as **lava.** As it cools, it forms underwater mountains or ridges. Sometimes these mountains rise above the surface of the water and form islands.

Plates can also slide past each other. They grind along faults, causing **earthquakes**.

> Underline what happens when ocean plates collide with one another.

> What causes earthquakes?
> _______________________
> _______________________

## FORCES ON EARTH'S SURFACE

As landforms are created, other forces work to wear them away. **Weathering** breaks larger rocks into smaller rocks. Changes in temperature can cause cracks in rocks. Water then gets into the cracks, expands as it freezes, and breaks the rocks. Rocks eventually break down into smaller pieces called sediment. Flowing water moves sediment to form new landforms, such as river deltas.

> Circle the three forces that can cause erosion.

Another force that wears down landforms is **erosion**. Erosion takes place when sediment is moved by ice, water, and wind.

## LANDFORMS INFLUENCE LIFE

Landforms influence where people live. For example, people might want to settle in an area with good soil and water. People change landforms in many ways. For example, engineers build tunnels through mountains to make roads. Farmers build terraces on steep hillsides.

## CHALLENGE ACTIVITY

**Critical Thinking: Drawing Inferences** Find out about a landform in your area that was changed by people. Write a report explaining why and how it was changed.

Name _________________________________ Class _______________ Date _____________

# Climate, Environment, and Resources

## CHAPTER SUMMARY

| CLIMATE | ENVIRONMENT | RESOURCES |
|---|---|---|
| front | deforestation | petroleum |
| rain shadow | savanna | solar energy |
| ocean currents | habitat | renewable resources |
| precipitation | desert | minerals |
| prevailing winds | ecosystem | fossil fuels |
| dry air | humus | soil |
| high pressure | rain forest | forest |
| hurricane | pollution | erosion |
| sunlight | extinct | hydroelectric power |
| Gulf Stream | river | water |

## COMPREHENSION AND CRITICAL THINKING

Use information from the graphic organizer to answer the following
questions.

1. **Explain** Choose a term from the *Environment* column that is affected by
   hydroelectric power. Write a sentence explaining how it is affected.

   _______________________________________________________________

   _______________________________________________________________

2. **Interpret** Under which other column could the word *desert* be listed? Why?

   _______________________________________________________________

   _______________________________________________________________

3. **Identify Cause and Effect** Explain how desertification occurs and why
   it is a problem.

   _______________________________________________________________

   _______________________________________________________________

   _______________________________________________________________

   _______________________________________________________________



          Interactive Reader and Study Guide

# Climate, Environment, and Resources

**Section 1**

**MAIN IDEAS**

1. While weather is short term, climate is a region's average weather over a long period.
2. The amount of sun at a given location is affected by Earth's tilt, movement, and shape.
3. Wind and water move heat around Earth, affecting how warm or wet a place is.
4. Mountains influence temperature and precipitation.

## Key Terms and Places

**weather** the short-term changes in the air for a given place and time

**climate** a region's average weather conditions over a long period

**prevailing winds** winds that blow in the same direction over large areas of Earth

**ocean currents** large streams of surface seawater

**front** a place where two air masses of different temperature or moisture content meet

## Section Summary

### UNDERSTANDING WEATHER AND CLIMATE

**Weather** is the condition of the atmosphere at a certain time and place. **Climate** is a region's average weather over a long time. Climate is affected mostly by two factors: sun and latitude. Energy from the sun falls more directly on the equator, so the hottest temperatures are near the equator. In general it gets colder as you move away from the equator to a higher latitude.

> **What are two important forces that affect climate?**
>
> ___________________________
>
> ___________________________

### SUN AND LOCATION

Heat from the sun moves around the Earth, partly through winds. Wind is caused by the rising and sinking of air. Cold air sinks, and warm air rises. At different latitudes winds tend to blow in the same direction. These **prevailing winds** can be from the west or east. Near the poles and in the subtropics, prevailing winds are easterlies. In the middle latitudes are the westerlies. Prevailing winds control an area's climate.

> **What causes wind?**
>
> ___________________________
>
> ___________________________



 Interactive Reader and Study Guide

## WIND AND WATER

Winds pick up moisture over oceans and dry out passing over land. At about 30° North and South latitude, dry winds cause many of the world's deserts.

**Ocean currents**—large streams of surface water—also move heat around. The Gulf Stream is a warm current that flows from the Gulf of Mexico across the Atlantic Ocean to western Europe.

Water heats and cools more slowly than land. Therefore, water helps to moderate the temperature of nearby land, keeping it from getting very hot or very cold.

> **Which heats and cools more slowly—land or water?**
> _________________________

A **front** is a place where two different air masses meet. In the United States and other regions, warm and cold air masses meet often, causing severe weather. These can include thunderstorms, blizzards, and tornadoes. Tornadoes are twisting funnels of air that touch the ground. Hurricanes are large tropical storms that form over water. They bring destructive high winds and heavy rain.

> **Which occurs in the tropics—tornadoes or hurricanes?**
> _________________________

## MOUNTAINS

Mountains also affect climate. Warm air blowing against a mountainside rises and cools. Clouds form, and precipitation falls on the side facing the wind. However, the air is dry by the time it goes over the mountain. This effect creates a rain shadow, a dry area on the side of the mountain facing land.

## CHALLENGE ACTIVITY

**Critical Thinking: Sequencing** Write a short description of the process leading up to the formation of a rain shadow. Draw and label a picture to go with your description.

# Climate, Environment, and Resources

### Section 2

**MAIN IDEAS**

1. Geographers use temperature, precipitation, and plant life to identify climate zones.
2. Tropical climates are wet and warm, while dry climates receive little or no rain.
3. Temperate climates have the most seasonal change.
4. Polar climates are cold and dry, while highland climates change with elevation.

## Key Terms and Places

**monsoon** winds that shift direction with the seasons and create wet and dry periods

**savanna** an area of tall grasses and scattered trees and shrubs

**steppe** a semi-dry grassland or prairie

**permafrost** permanently frozen layers of soil

## Section Summary

### MAJOR CLIMATE ZONES

We can divide Earth into five climate zones: tropical, temperate, polar, dry, and highland. Tropical climates appear near the equator, temperate climates are found in the middle latitudes, and polar climates occur near the poles. Dry and highland climates can appear at different latitudes.

> Underline the names of the five climate zones.

### TROPICAL AND DRY CLIMATES

Humid tropical climates occur near the equator. Some are warm and rainy throughout the year. Others have **monsoons**—winds that shift directions and create wet and dry seasons. Rain forests need a humid climate to thrive and support thousands of species.

Moving away from the equator, we find tropical savanna climates. A **savanna** is an area of tall grasses and scattered trees and shrubs. A long, hot dry season is followed by short periods of rain.

> What happens when monsoon winds change direction?
>
> _______________________
>
> _______________________

Deserts are hot and dry. At night, the dry air cools quickly; desert nights can be cold. Only a few living things survive in a desert. Sometimes **steppes**—dry grasslands—are found near deserts.

## TEMPERATE CLIMATES

Away from the ocean in the middle latitudes are humid continental climates. Most have four distinct seasons, with hot summers and cold winters. In this climate, weather often changes quickly when cold and warm air masses meet.

> Underline the name of the climate that can have four distinct seasons.

A Mediterranean climate has hot, sunny summers and mild, wet winters. They occur near the ocean, and the climate is mostly pleasant. People like to vacation in these climates. Only small, scattered trees survive in these areas.

> What do people typically like to do in Mediterranean climates?
> ___________________________

East coasts near the tropics have humid subtropical climates, because of winds bringing in moisture from the ocean. They have hot, wet summers and mild winters. Marine west coast climates occur farther north and also get moisture from prevailing winds coming in from the ocean.

## POLAR AND HIGHLAND CLIMATES

Subarctic climate occurs south of the Arctic Ocean. Winters are long and cold, and summers are cool. There is enough precipitation to support forests. At the same latitude near the coasts, tundra climate is also cold, but too dry for trees to survive. In parts of the tundra, soil is frozen as **permafrost**.

> Can there be forests in subarctic climates?
> ___________________________

Ice cap climates are the coldest on Earth. There is little precipitation and little vegetation.

Highland, or mountain, climate changes with elevation. As you go up a mountain, the climate may go from tropical to polar.

## CHALLENGE ACTIVITY

**Critical Thinking: Comparing and Contrasting** Create a table showing the differences and similarities between any two types of climate.

# Climate, Environment, and Resources

## Section 3

**MAIN IDEAS**
1. The environment and life are interconnected and exist in a fragile balance.
2. Soils play an important role in the environment.

## Key Terms and Places

**environment**  a plant or animal's surroundings

**ecosystem**  any place where plants and animals depend upon each other and their environment for survival

**habitat**  the place where a plant or animal lives

**extinct**  to die out completely

**humus**  decayed plant or animal matter

**desertification**  the slow process of losing soil fertility and plant life

## Section Summary

### THE ENVIRONMENT AND LIFE

Plants and animals cannot live just anywhere. They must have the right surroundings, or **environment**. Climate, land features, and water are all part of a living thing's environment. If an area has everything a living thing needs, it can be a **habitat** for that species.

Many plants and animals usually share a habitat. Small animals eat plants, and then large animals eat the small animals. Species are connected in many ways. A community of connected species is called an **ecosystem**. Ecosystems can be as small as a pond or as large as the entire Earth.

> **How large can an ecosystem be?**

Geographers study how changes in environments affect living things. Natural events and human actions change environments. Natural events include forest fires, disease, and climate changes. Human actions include clearing land and polluting.

> **Underline the human actions that can cause changes in an environment.**

If a change to the environment is extreme, a species might become **extinct**, or die out completely.

## SOILS AND THE ENVIRONMENT

Without soil, much of our food would not exist. Soil forms in layers over hundreds or thousands of years. The most fertile layer, the topsoil, has the most humus. **Humus** is decayed plant or animal matter.

The next layer, the subsoil, has less humus and more material from rocks. Soil gets minerals from these rocks. Below the subsoil is mostly rock.

An environment's soil affects which plants can grow there. Fertile soils have lots of humus and minerals. Fertile soils also need to contain water and small air spaces.

Soils can lose fertility from erosion by wind or water. Soil can also lose fertility from planting the same crops repeatedly. If soil becomes worn out and can no longer support plants, **desertification** can occur.

## CHALLENGE ACTIVITY

**Critical Thinking: Drawing Inferences** Consider the interconnections in your environment. As you go through a normal day, keep a list of the sources you rely on for energy, food, and water.

> **What does it mean to become extinct?**
> _______________________
> _______________________

> **Which has more humus—topsoil or subsoil?**
> _______________________

> **Underline two things found in fertile soil.**

# Climate, Environment, and Resources

**Section 4**

**MAIN IDEAS**

1. Earth provides valuable resources for our use.
2. Energy resources provide fuel, heat, and electricity.
3. Mineral resources include metals, rocks, and salt.
4. Resources shape people's lives and countries' wealth.

## Key Terms and Places

**natural resource**  any material in nature that people use and value

**renewable resources**  resources that can be replaced naturally

**nonrenewable resources**  resources that cannot be replaced

**deforestation**  the loss of forestland

**reforestation**  planting trees to replace lost forestland

**fossil fuels**  nonrenewable resources formed from the remains of ancient plants and animals

**hydroelectric power**  the production of electricity by moving water

## Section Summary

### EARTH'S VALUABLE RESOURCES

Anything in nature that people use and value is a **natural resource**. These include such ordinary things as air, water, and soil. Resources such as trees are called **renewable resources** because Earth replaces them naturally. Those that cannot be replaced, such as oil, are called **nonrenewable resources**.

Air and water are renewable resources, but pollution can damage both. Some people get their water from underground wells, which can run out if too many people use them.

> Are air and water renewable or nonrenewable resources?
>
> _______________________

Soil is needed for all plant growth, including trees in forests. We get lumber, medicine, nuts, and rubber from forests. Soil and trees are renewable, but must be protected. The loss of forests is called **deforestation**. When we plant trees to replace lost forests, we call it **reforestation**.

> Underline the resources we can get from a forest.

**Section 4, *continued***

## ENERGY RESOURCES

Most of our energy comes from **fossil fuels,** which
are formed from the remains of ancient living things.
These include coal, oil, and natural gas.

We use coal mostly for electricity, but it causes
air pollution. An advantage of coal is that Earth still
has a large supply. Another fossil fuel is petroleum,
or oil. It is used to make gasoline and heating oil.
Oil can be turned into plastics and other products.
Oil also causes pollution, but we depend on it for
much of our energy. The cleanest fossil fuel is natural
gas, which is used mainly for cooking and heating.

Renewable energy resources include
**hydroelectric power**—the creation of electricity
by moving water. This is accomplished mainly by
building dams on rivers. Other renewable energy
sources are wind, solar, and nuclear energy. Nuclear
energy produces dangerous waste material that
must be stored for thousands of years.

> **Where does gasoline come from?**
> _______________________

> **What is the cleanest-burning fossil fuel?**
> _______________________

## MINERAL RESOURCES

Minerals are solid substances in the Earth's crust
formed from nonliving matter. Like fossil fuels,
minerals are nonrenewable. Types of minerals include
metals, rocks and gemstones, and salt. Mineral uses
include making steel from iron, making window
glass from quartz, and using stone as a building
material. We also use minerals to make jewelry,
coins, and many other common objects.

> **Name two uses of minerals.**
> _______________________
> _______________________

## RESOURCES AND PEOPLE

Some places are rich in natural resources. Resources
such as fertile farmland, forests, and oil have helped
the United States become a powerful country with a
strong economy.

## CHALLENGE ACTIVITY

**Critical Thinking: Drawing Inferences** Write a short essay explaining
why we still use coal, even though it causes pollution.

# The World's People

## CHAPTER SUMMARY

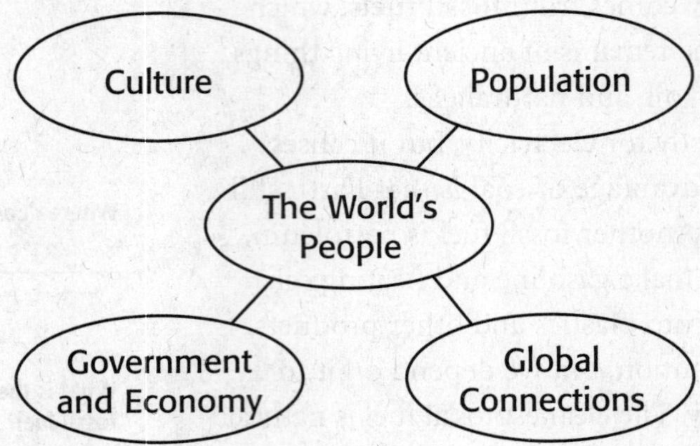


## COMPREHENSION AND CRITICAL THINKING

Use information from the graphic organizer to answer the following questions.

**1. Identify** What four types of government would you list in the "Government and Economy" circle to make the graphic organizer more detailed?

_______________________________________________

**2. Analyze** Give an example of how the population in an area can affect the economy.

_______________________________________________

_______________________________________________

**3. Make an Inference** How can global connections affect the culture in an area?

_______________________________________________

_______________________________________________

**4. Draw a Conclusion** If the world's population continues to grow rapidly, what might be the most important challenges for people to overcome?

_______________________________________________

_______________________________________________

_______________________________________________

# The World's People

**Section 1**

**MAIN IDEAS**

1. Culture is the set of beliefs, goals, and practices that a group of people share.
2. The world includes many different culture groups.
3. New ideas and events lead to changes in culture.

## Key Terms

**culture**  the set of beliefs, values, and practices a group of people have in common
**culture trait**  an activity or behavior in which people often take part
**culture region**  an area in which people have many shared culture traits
**ethnic group**  a group of people who share a common culture and ancestry
**cultural diversity**  having a variety of cultures in the same area
**cultural diffusion**  the spread of culture traits from one region to another

## Section Summary

### WHAT IS CULTURE?

**Culture** is the set of beliefs, values, and practices a group of people have in common. Everything in day-to-day life is part of culture, including language, religion, clothes, music, and foods. People everywhere share certain basic cultural features, such as forming a government, educating children, and creating art or music. However, people practice these things in different ways, making each culture unique.

> Underline the sentence which lists some parts of culture.

> What are some ways cultures develop?
> ___________________________
> ___________________________

**Culture traits** are activities or behaviors in which people often take part, such as language and popular sports. People share some culture traits, but not others. For example, people eat using forks, chopsticks, or their fingers in different areas.

### CULTURE GROUPS

There are thousands of different cultures in the world. People who share a culture are part of a culture group that may be based on things like age or religion.

## Section 1, *continued*

A **culture region** is an area in which people have many shared culture traits such as language, religion, or lifestyle. One large culture region is the Arab world in Southwest Asia and North Africa. A country may have several different culture regions, or just a single region, such as Japan.

A culture region may be based on an **ethnic group**, a group of people who share the same religion, traditions, language, or foods. **Cultural diversity** is having a variety of cultures in the same area. It can create a variety of ideas and practices, but it can also lead to conflict.

**How can cultural diversity affect the people in an area?**

_______________________________

_______________________________

## CHANGES IN CULTURE

Cultures are constantly changing. They can change through the development of new ideas or contact with other societies. New ideas such as the development of electricity, motion pictures, and the Internet have changed what people do and how they communicate. When two cultures come in close contact, both usually change. For example, the Spanish and Native American cultures changed when the Spanish conquered the Americas.

**Underline the sentences that describe how cultures change.**

**Cultural diffusion** is the spread of culture traits from one part of the world to another. It can happen when people move and bring their culture with them. New ideas and customs, such as baseball or clothing styles, can spread from one place to another as people learn about them.

**What are two ways cultural diffusion occurs?**

_______________________________

## CHALLENGE ACTIVITY

**Critical Thinking: Drawing Inferences** Consider all of the parts of your culture that have been influenced by other cultures. During a normal day, keep a list of all the things you use or do that you think have been influenced by other cultures.

# The World's People

**Section 2**

**MAIN IDEAS**

1. The study of population patterns helps geographers learn about the world.
2. Population statistics and trends are important measures of population change.

## Key Terms

**population** the total number of people in a given area

**population density** a measure of the number of people living in an area, usually expressed as persons per square mile or square kilometer

**birthrate** the annual number of births per 1,000 people

**migration** the process of moving from one place to live in another

## Section Summary

### POPULATION PATTERNS

**Population** is the total number of people in a given area. Geographers study population patterns to learn about the world.

Some places are crowded with people, while others are almost empty. **Population density** is a measure of the number of people living in an area, usually expressed as persons per square mile or square kilometer. It describes how crowded a place is, which in turn affects how people live. In places with a high density, there is little open space, buildings are taller, and roads are more crowded than places with lower density. They also often have more products available for a variety of shoppers.

High density areas often have fertile soil, available water, and a favorable climate for agriculture. Areas that are less dense often have harsh land or climate that makes survival harder.

> Underline the two sentences that describe the effects of population density on a place.

> What is the land and climate often like in areas of high population density?
>
> _______________________________
>
> _______________________________

### POPULATION CHANGE

The number of people living in an area affects jobs, housing, schools, medical care, available food, and many other things. Geographers study population



    Interactive Reader and Study Guide

**Section 2, continued**

changes and world trends to understand how people live.

Three statistics are important to study a country's population over time. **Birthrate** is the annual number of births per 1,000 people. Death rate is the annual number of deaths per 1,000 people. The rate of natural increase is found by subtracting the death rate from the birthrate.

Some areas have low rates of natural increase, such as Europe and North America. Some countries in Africa and Asia have very high rates of natural increase. High rates make it hard for countries to develop economically because they need to provide jobs, education, and medical care for a growing population.

**Migration** is the process of moving from one place to live in another. People may leave a place because of problems there, such as war, famine, drought, or lack of jobs. Other people may move to find political or religious freedom or economic opportunities in a new place.

The world's population has grown very rapidly in the last 200 years. Better health care and food supplies have helped more babies survive and eventually have children of their own. Many industrialized countries currently have slow population growth while other countries have very fast growth. Fast growth puts a strain on resources, housing, and government aid.

> Underline the sentence that tells how to calculate the rate of natural increase.

> Why do high rates of natural increase make it hard for a country to develop economically?
> _______________________________
> _______________________________

> How has the world's population changed during the last 200 years?
> _______________________________
> _______________________________

## CHALLENGE ACTIVITY

**Critical Thinking: Identify Cause and Effect** Find out the population density of your city or town. Write down ways that this density affects your life and the lives of others.

# The World's People

**Section 3**

**MAIN IDEAS**

1. The governments of the world include democracy, monarchy, dictatorship, and communism.
2. Different economic activities and systems exist throughout the world.
3. Geographers group the countries of the world based on their level of economic development.

## Key Terms

**democracy** a form of government in which the people elect leaders and rule by majority

**communism** a political system in which the government owns all property and dominates all aspects of life in the country

**market economy** a system based on private ownership, free trade, and competition

**command economy** a system in which the central government makes all economic decisions

**gross domestic product (GDP)** the value of all goods and services produced within a country in a single year

**developed countries** countries with strong economies and a high quality of life

**developing countries** countries with less productive economies and a lower quality of life

## Section Summary

### GOVERNMENTS OF THE WORLD

People form governments to make laws, regulate business, and provide aid to people. A **democracy** is a form of government in which the people elect leaders and rule by majority. Most democracies protect people's rights to freedom of speech, religion, and a free press.

> **What rights are protected in most democracies?**
> _________________________________
> _________________________________

Monarchies are ruled directly by a king or queen who holds all the power. Dictatorships are also ruled by a single person. Dictators hold all the power and often rule by force. **Communism** is a system in which the government owns all property and dominates all aspects of life in the country. In most Communist states, the people have restricted rights and little freedom.

> **What types of government are ruled by a single person?**
> _________________________________
> _________________________________



  Interactive Reader and Study Guide

## ECONOMIES OF THE WORLD

Primary industries provide natural resources to others through work such as farming, fishing, and mining. Secondary industries use raw materials to manufacture other products such as automobiles or furniture. Tertiary industries exchange goods and services through retail stores, health care and educational organizations, and so on. Quaternary industries involve workers such as architects, lawyers, and scientists who research and distribute information.

In a traditional economy, people make and use their own goods with little exchange of goods. A **market economy** is based on free trade and competition. People buy and sell as they wish and prices are determined by supply and demand. In a **command economy**, the government decides what to produce and what prices will be.

> **What are the four levels of economic activity?**
> _______________________
> _______________________

> **Underline the sentence that describes who controls a command economy.**

## ECONOMIC DEVELOPMENT

One measure of economic development is **gross domestic product (GDP)**, the value of all goods and services produced within a country in a single year. Other ways include the level of industrialization and the quality of life.

**Developed countries** have strong, wealthy economies and high standards of living. **Developing countries** have poorer economies and a lower quality of life. About two thirds of the world's people live in developing countries with poor education and little access to health care or telecommunications.

> **Which type of country is more industrialized?**
> _______________________
> _______________________

## CHALLENGE ACTIVITY

**Critical Thinking: Classify** Select a family member or friend and ask them about their job. Classify it as one of the four basic types and describe why it is important in our society.

**Section 4**

**MAIN IDEAS**

1. Globalization links the world's countries together through culture and trade.
2. The world community works together to solve global conflicts and crises.

# Key Terms

**globalization** the process in which countries are increasingly linked to each other through culture and trade

**popular culture** culture traits that are well known and widely accepted

**interdependence** the reliance of one country on the resources, goods, or services of another country

**United Nations (UN)** an organization of the world's countries that promotes peace and security around the globe

**humanitarian aid** assistance to people in distress

# Section Summary

## GLOBALIZATION

People around the world are more closely linked than ever before. **Globalization** is the process in which countries are increasingly linked to each other through culture and trade. Improvements in technology and communication have increased globalization.

**Popular culture** consists of culture traits that are well known and widely accepted. These traits can include food, sports, music, and movies. The United States has a great influence on popular culture through sales of American products and the use of English for business, science, and education around the world. It is also greatly influenced by other countries.

World businesses are connected through trade. Companies may make products in many different countries or use products from around the world. **Interdependence** occurs when countries depend on each other for resources, goods, or services.

> **Underline the sentence which describes two ways countries are linked together.**

> **What are four traits that can be considered part of popular culture?**
>
> ___________________________
>
> ___________________________

## Section 4, *continued*

Companies and consumers depend on goods produced elsewhere.

> **What two groups might depend on goods produced elsewhere?**
>
> _______________________________
>
> _______________________________

## A WORLD COMMUNITY

Because places around the world are connected closely, what happens in one place affects others. The world community works together to promote cooperation between countries.

When conflicts occur, countries from around the world try to settle them. The **United Nations (UN)** is an association of nearly 200 countries dedicated to promoting peace and security.

> **Underline the sentence that describes the main goals of the United Nations.**

Crises such as earthquakes, floods, drought, or a tsunami can leave people in great need. Groups from around the world provide **humanitarian aid**, or assistance to people in distress. Some groups help refugees or provide medical care.

## CHALLENGE ACTIVITY

**Critical Thinking: Contrast** Talk to a parent or other adult about their knowledge of other countries and their connections to them when they were young. Write a short essay that contrasts their global connections with yours.





Interactive Reader and Study Guide

# Early History of the Americas

## CHAPTER SUMMARY

### Comparing the Maya, Aztec, and Inca Civilizations

| | Maya | Aztec | Inca |
|---|---|---|---|
| **Taxes** | | Tributes of cotton, gold, or food | Labor |
| **Upper Classes** | King, priests, warriors, merchants | King, nobles, priests, warriors | |
| **Lower Classes** | Farmers | Farmers and laborers; slaves | |
| **Language** | Writing | | No writing |
| **Sacrifices** | Blood, human sacrifice | | Llamas, cloth, or food |
| **Structures** | | Temples, palace, causeways | Massive stones in buildings, network of roads |
| **Condition of Empire when Spanish Arrived** | Civilization in decline | Empire at peak | |

## COMPREHENSION AND CRITICAL THINKING

Use the answers to the following questions to fill in the graphic organizer above.

**1. Contrast** In what form were taxes collected in each civilization?

_______________________________________________________________

_______________________________________________________________

**2. Compare and Contrast** Name at least two differences and two similarities among the Maya, Aztec, and Inca civilizations.

_______________________________________________________________

_______________________________________________________________

**3. Draw a Conclusion** Although they were greatly outnumbered, Spanish conquistadors conquered all three empires. Explain how this was possible.

_______________________________________________________________

_______________________________________________________________

# Early History of the Americas

**Section 1**

**MAIN IDEAS**

1. Geography helped shape the lives of the early Maya.
2. During the Classic Age, the Maya built great cities linked by trade.
3. Maya culture included a strict social structure, a religion with many gods, and achievements in science and the arts.
4. The decline of Maya civilization began in the 900s.

## Key Terms and Places

**maize** corn

**Palenque** Maya city in which the king Pacal had a temple built to record his achievements

**observatories** buildings from which people could study the sky

## Section Summary

### GEOGRAPHY AND THE EARLY MAYA

Mesoamerica extends from central Mexico to the northern part of Central America. The Maya (MY-uh) civilization developed here around 1000 BC. Thick forests covered the area, so the Maya cleared the areas to farm. They grew **maize**, or corn, and beans, squash, and avocados. The forest was also a source of many resources, including animals for food and wood for building materials. The Maya lived in villages. By AD 200, the Maya were building large cities.

> Underline the description of the land included in the area called Mesoamerica.

### THE CLASSIC AGE

Maya civilization was at its peak between AD 250 and 900, a period called the Classic Age. There were more than 40 Maya cities. They traded crops, wood, jade, and obsidian.

The Maya built large stone pyramids, temples, and palaces. Some buildings honored local kings. A temple built in the city of **Palenque** (pah-LENG-kay) honored the king Pacal (puh-KAHL). The Maya built canals to bring water to the cities. They also shaped hillsides into flat terraces for crops.

> List three valued Maya exports.
> _______________________
> _______________________



   Interactive Reader and Study Guide

## MAYA CULTURE

The Maya had a complex social structure. Kings held the highest position. Priests, warriors, and merchants made up the upper class. Most Maya belonged to lower class farming families. Maya farmers had to "pay" the rulers with some of their crops and with goods such as cloth and salt. They also had to help build temples and other buildings.

The Maya worshipped many gods. Each god represented a different aspect of life. The Maya tried to keep the gods happy by giving them blood.

Maya achievements in art, architecture, math, science, and writing were remarkable. They built **observatories** for priests to study the stars. They developed a calendar that had 365 days. The Maya developed a number system and a writing system. They also made jade and gold jewelry.

> **What did the Maya think their gods wanted?**
> _______________________

## DECLINE OF MAYA CIVILIZATION

Maya civilization began to collapse in the 900s. They stopped building large buildings and left the cities for the countryside. Historians are not sure why this happened, but there are several theories.

Some historians believe that Maya farmers kept planting the same crop over and over, which weakened the soil. This may have caused more competition and war between the cities. The people may have decided to rebel against their kings' demands. There probably were many factors that led to the decline of the Maya civilization.

> **List two factors that may have contributed to the decline of the Maya civilization.**
> _______________________
> _______________________
> _______________________

## CHALLENGE ACTIVITY

**Critical Thinking: Drawing Inferences** One source of information about the Maya comes from kings like Pacal, who dedicated a temple to his achievements. Draw a building that honors our culture. Include details that would help future historians reconstruct 21st century life.

# Early History of the Americas

**Section 2**

**MAIN IDEAS**

1. The Aztecs built a rich and powerful empire in central Mexico.
2. Social structure, religion, and warfare shaped life in the empire.
3. Hernán Cortés conquered the Aztec Empire in 1521.

## Key Terms and Places

**Tenochtitlán** island city and capital of the Aztec Empire

**causeways** raised roads across water or wet ground

**conquistadors** Spanish conquerors

## Section Summary

### THE AZTECS BUILD AN EMPIRE

The first Aztecs were poor farmers who migrated south to central Mexico. Other tribes had taken the good farmland, so the Aztecs settled on a swampy island in Lake Texcoco (tays-KOH-koh). In 1325, they began building their capital here.

War was key to the Aztecs' rise to power. The Aztec warriors conquered many towns and made the conquered people pay tributes of cotton, gold, or food. The Aztecs also controlled a large trade network.

The Aztecs' power and wealth was most visible in the capital, **Tenochtitlán** (tay-NAWCH-teet-LAHN). The Aztecs built three **causeways** to connect the island city to the lakeshore. At its peak, Tenochtitlán had about 200,000 people. The city had temples, a palace, and a busy market.

> **Where were the humble origins of the mighty Aztecs?**
> _______________________
> _______________________
> _______________________

> **What were the two key ways that the Aztecs became rich?**
> _______________________
> _______________________
> _______________________

### LIFE IN THE EMPIRE

Aztec society had clearly defined social classes. The king was the most important person. He was in charge of law, trade, tribute, and warfare. The nobles, including tax collectors and judges, helped the king with his duties. Below the king and nobles were priests and warriors. Priests had great

> **Underline the sentence that describes the responsibilities of the king.**



 Interactive Reader and Study Guide

**Section 2,** *continued*

influence over Aztecs. Warriors were highly respected. Below priests and warriors were merchants and artisans, and then farmers and laborers. Slaves were lowest in society.

The Aztecs believed that gods ruled all parts of life and sacrifice was necessary to keep the gods happy. In rituals priests cut open victims' chests to give blood to the gods and sacrificed nearly 10,000 humans a year.

The Aztecs studied astronomy. Their calendar was much like the Maya calendar. The Aztecs had a rich artistic tradition and their own writing system. They also had a strong oral tradition.

**Why were Aztec religious ceremonies so bloody?**

_______________________

_______________________

_______________________

_______________________

## CORTÉS CONQUERS THE AZTECS

In 1519 Hernán Cortés (er-NAHN kawr-TEZ) led a group of Spanish conquerors called **conquistadors** into Mexico. Their motives were to seek gold, claim land, and spread their religion. The Aztec ruler, Moctezuma II (MAWK-tay-SOO-mah), thought Cortés was a god. Moctezuma gave Cortés many gifts, including gold. Wanting more gold, Cortés took Moctezuma prisoner. Enraged, the Aztecs attacked the Spanish. They drove the Spanish out of the city, but Moctezuma was killed.

To defeat the Aztecs, the Spanish allied with people who resented the Aztec rulers. Together they used guns and rode horses. The Spanish also carried diseases like smallpox that killed many Aztecs. In 1521 the Spanish brought the Aztec Empire to an end.

**What factors contributed to the Aztecs' defeat?**

_______________________

_______________________

_______________________

_______________________

## CHALLENGE ACTIVITY

**Critical Thinking: Drawing Inferences** What do you think about Hernán Cortés and his actions toward the Aztecs? Write a one-page paper defending your opinion. Give examples to support your opinion.

# Early History of the Americas

**Section 3**

**MAIN IDEAS**

1. The Incas created an empire with a strong central government in South America.
2. Life in the Inca Empire was influenced by social structure, religion, and the Incas' cultural achievements.
3. Francisco Pizarro conquered the Incas and took control of the region in 1537.

## Key Terms and Places

**Cuzco** capital of the Inca Empire, located in present-day Peru

**Quechua** the official language of the Incas

**masonry** stonework

## Section Summary

### THE INCAS CREATE AN EMPIRE

While the Aztecs were ruling Mexico, the Incas were building an empire in South America. The Incas began as a small tribe high in the Andes. They built their capital, **Cuzco**, in modern-day Peru. In the mid-1400s, the ruler Pachacuti (pah-chah-KOO-tee) led the Incas to expand their territory. By the early 1500s, the Inca Empire stretched from Ecuador to central Chile.

> On what continent did the Incas build their empire?
> ________________________
> ________________________

> Where did the Inca tribe originate?
> ________________________
> ________________________

To rule this empire of 12 million people, the Incas formed a strong central government. The Incas replaced local leaders of conquered areas with new people loyal to the Inca government. The Incas established an official language, **Quechua** (KE-chuh-wuh). The Incas paid taxes in the form of labor. This labor tax system was called the *mita* (MEE-tah). There were no merchants or markets. Instead, government officials would distribute goods collected through the *mita*.

> How did Incas get food, clothing and other goods?
> ________________________
> ________________________

### LIFE IN THE INCA EMPIRE

Inca society had two main social classes. The emperor, priests, and government officials were the



                    Interactive Reader and Study Guide

upper class. The upper class lived in Cuzco and did not pay the labor tax. The lower class included farmers, artisans, and servants. Most Incas were farmers. They could not own more goods than they needed.

The Inca religion was based on the belief that Inca rulers were related to the sun god and never really died. Inca ceremonies often included sacrifice of llamas, cloth, or food. They also believed certain natural landforms had magical powers.

> **The Incas believed their rulers were related to whom?**
>
> _______________________
>
> _______________________

Incas are known for their **masonry**, or stonework. They built massive buildings and a network of roads. Inca artisans made beautiful pottery, jewelry, and textiles. The Incas had no written language. Instead, they kept records with cords and passed down stories and songs orally.

> **The Incas excelled in the use of what building material?**
>
> _______________________
>
> _______________________

## PIZARRO CONQUERS THE INCAS

In the late 1520s a civil war began between an Inca ruler's two sons, Atahualpa (ah-tah-WAHL-pah) and Huáscar (WAHS-kahr). Atahualpa won, but the war had weakened the Inca army. On his way to be crowned king, Atahualpa heard that conquistadors led by Francisco Pizarro had arrived in the Inca Empire. When Atahualpa came to meet them, the Spanish captured him. They attacked and killed thousands of Inca soldiers. The Incas brought gold and silver to offer for Atahualpa's return. But instead the Spanish killed him. The Spanish defeated the Incas and ruled their lands for the next 300 years.

## CHALLENGE ACTIVITY

**Critical Thinking: Drawing Inferences** The Incas used labor as a form of currency. What are the advantages and disadvantages of this type of economic system? Write a brief essay explaining your answer.

# Mexico

## CHAPTER SUMMARY

- UNEMPLOYMENT
- OIL
- MAQUILADORAS
- CENTRAL PLATEAU
- MEXICAN INDIAN HERITAGE
- **MEXICO**
- SCENIC LANDSCAPES
- SILVER
- SPANISH COLONIAL HERITAGE
- SUBSISTENCE FARMING
- NORTHERN DESERT

## COMPREHENSION AND CRITICAL THINKING

Use information from the graphic organizer to answer the following questions.

**1. Identify**  Which three of the terms above are key natural resources?

_______________________________________________________________

_______________________________________________________________

**2. Analyze**  Which item has to do with physical geography, history, and economics? Explain your answer.

_______________________________________________________________

_______________________________________________________________

**3. Interpret**  What six items describe Mexico's economy today?

_______________________________________________________________

_______________________________________________________________

_______________________________________________________________

_______________________________________________________________

# Mexico

**Section 1**

**MAIN IDEAS**

1. Mexico's physical features include plateaus, mountains, and coastal lowlands.
2. Mexico's climate and vegetation include deserts, tropical forests, and cool highlands.
3. Key natural resources in Mexico include oil, silver, gold, and scenic landscapes.

## Key Terms and Places

**Río Bravo**  Rio Grande, forms part of Mexico's border with the U.S.

**peninsula**  piece of land surrounded on three sides by water

**Baja California**  peninsula stretching from northern Mexico into the Pacific Ocean

**Gulf of Mexico**  body of water that forms Mexico's eastern border

**Yucatán Peninsula**  land separating the Gulf of Mexico from the Caribbean Sea

**Sierra Madre**  "mother range" made up of three mountain ranges in Mexico

## Section Summary

### PHYSICAL FEATURES

Mexico shares a long border with the United States. Part of this border is formed by a river called the **Río Bravo,** known as the Rio Grande in the United States. Mexico's western border is the Pacific Ocean, where a long **peninsula** called **Baja California** stretches south from northern Mexico. In the east, the **Yucatán Peninsula** separates the **Gulf of Mexico** from the Caribbean Sea.

Underline the names of Mexico's two peninsulas.

The interior of Mexico is mostly the high, rugged Mexican Plateau, which rises in the west to the Sierra Madre Occidental. In the east it meets the Sierra Madre Oriental. **Sierra Madre** means "mother range." The country's capital, Mexico City, lies at the southern end of the plateau in the Valley of Mexico. The city has earthquakes, and to the south there are active volcanoes.

Where is Mexico City located?

____________________________

____________________________

From the central highlands, the land slopes down to Mexico's sunny beaches. In the east the Gulf coastal plain is wide, and there are many farms.





Interactive Reader and Study Guide

**Section 1, *continued***

The Yucatán Peninsula is mostly flat. The limestone rock there has eroded to form caves and steep depressions called sinkholes, many of which are filled with water.

## CLIMATE AND VEGETATION

Mexico has many climates with different types of vegetation. The mountains and plateaus are cool, and freezing temperatures can reach all the way to Mexico City. The mountain valleys are mild, and the southern coast is also pleasant. Summer rains support tropical rain forests, where animals such as jaguars, monkeys, and anteaters live. The Yucatán Peninsula is hot and dry, supporting only scrub forests. The north is also dry, much of it covered by the Sonoran and Chihuahuan deserts.

## NATURAL RESOURCES

Oil is an important resource. Mexico sells a lot of oil to the United States. Before oil was discovered, minerals were the most valuable resource. Today Mexico mines more silver than any other country. Copper, lead, gold, and zinc are also mined.

Another important resource is water. Unfortunately, this resource is scarce in parts of Mexico, especially the north. However, the water surrounding Mexico draws many tourists to the country's scenic beaches.

## CHALLENGE ACTIVITY

**Critical Thinking: Making Predictions** Write a paragraph making a prediction about which of Mexico's resources will be most important in Mexico's future. Support your prediction with information you learned in the section.

> How is the terrain in the Yucatán Peninsula different from most of Mexico?

> Underline the deserts in the north of Mexico.

> What is Mexico's most important mineral product?

# Mexico

**Section 2**

**MAIN IDEAS**

**1.** Early cultures of Mexico included the Olmec, the Maya, and the Aztec.

**2.** Mexico's period as a Spanish colony and its struggles since independence have shaped its culture.

**3.** Spanish and native cultures have influenced Mexico's customs and traditions today.

## Key Terms and Places

**empire** a land with different territories and peoples under a single ruler

**mestizos** the Spanish name for people of mixed European and Indian ancestry

**missions** church outposts

**haciendas** huge expanses of farm or ranch land

## Section Summary

### EARLY CULTURES

People grew corn, beans, and squash in Mexico as early as 5,000 years ago. About 1500 BC the Olmec settled on the southern coast of the Gulf of Mexico. They built temples and statues. About AD 250 the Maya built cities in Mexico and Central America. They were astronomers and left written records. Maya civilization collapsed after AD 900.

> Underline the achievements of the Olmec and Maya.

Later, the Aztecs moved into central Mexico. In 1325 they founded their capital, Tenochtitlán. They built an **empire** through conquest of other tribes.

### COLONIAL MEXICO AND INDEPENDENCE

In 1519 a Spanish soldier, Hernán Cortés, arrived in Mexico with guns, horses, and about 600 soldiers. The Spanish also brought diseases, which hurt the Aztec. This helped Cortés defeat the Aztec in 1521. He called the land New Spain.

> How long did it take Cortés to conquer the Aztec?
>
> _________________________________



     Interactive Reader and Study Guide

Many people in New Spain were of mixed European and Indian ancestry and were called **mestizos**. The Catholic Church was important in the colony. Priests tried to convert the Indians, traveling far north to build **missions**.

Spain was eager to mine gold and silver in Mexico. The native people and enslaved Africans did most of the mining. They also worked the huge farms and ranches, called **haciendas**, that were owned by people of Spanish ancestry.

Mexico gained independence in 1821. Miguel Hidalgo started the revolt by asking for equality in 1810. Later, Texas broke away from Mexico and joined the United States. The two countries fought over its border in the Mexican War. Mexico lost the war and almost half its territory.

In the mid-1800s, the popular president Benito Juárez made many reforms. But in the early 1900s the government helped the hacienda owners take land from the peasants. People were angry and started the Mexican Revolution in 1910. In 1920 a new government took land from the large landowners and gave it back to the peasants.

> Who owned the haciendas?
> ______________________

> What did the government do that made people angry?
> ______________________
> ______________________

## CULTURE

In Mexico language is tied to ethnic groups. Speaking an American Indian language identifies a person as Indian. Mexicans have combined Indian religious practices with Catholic practices. One example is the holiday for remembering ancestors, called Day of the Dead. It follows native traditions, but is celebrated on November 1 and 2—the same dates as similar Catholic holidays.

> Underline the Indian aspect of Day of the Dead.

## CHALLENGE ACTIVITY

**Critical Thinking: Sequencing** Make a time line with important dates and events in Mexican history.



 Interactive Reader and Study Guide

# Mexico

**Section 3**

---

**MAIN IDEAS**

1. Government has traditionally played a large role in Mexico's economy.
2. Mexico has four distinct culture regions.

---

## Key Terms and Places

**inflation**  a rise in prices that occurs when currency loses its buying power

**slash-and-burn agriculture**  the practice of burning forest to clear land for planting

**cash crop**  a crop that farmers grow mainly to sell for a profit

**Mexico City**  the world's second-largest city and Mexico's capital

**smog**  a mixture of smoke, chemicals, and fog

**maquiladoras**  U.S.- and foreign-owned factories in Mexico

## Section Summary

### GOVERNMENT AND ECONOMY

Although Mexico is a democracy, one political party ran the government for 71 years. This ended in 2000 when Vicente Fox was elected president. Like other developing countries, Mexico has foreign debts, unemployment, and **inflation**. Due to the North American Free Trade Agreement (NAFTA), Mexico now sells more products to its neighbors. Trucks bring **cash crops** like fruits and vegetables to the United States. Some farmers who don't own much land and only grow enough to feed their families use **slash-and-burn agriculture**.

Mexicans work in oil fields and in factories. Many Mexicans also come to the U.S. looking for work. Tourists visit Mexico to enjoy its attractions.

What is a developing country?

_______________________

How did NAFTA change trade for Mexico?

_______________________

### MEXICO'S CULTURE REGIONS

Mexico has four culture regions that differ from one another in population, resources, and climate.



  Interactive Reader and Study Guide

**Section 3, *continued***

The Greater **Mexico City** region includes the capital and about 50 nearby cities. More than 19 million people make Mexico City the world's second-largest city. Many people move there to look for work, and air pollution has become a problem. The mountains trap the **smog**—a mixture of smoke, chemicals, and fog. Poverty is also a problem.

Many cities in Mexico's central region were colonial mining or ranching centers. Mexico's colonial heritage can be seen today in the churches and public squares of this region. Family farmers grow vegetables and corn in the fertile valleys. In recent years, cities such as Guadalajara have attracted new industries from Mexico City.

> Underline information about the current economy of Mexico's central region. Circle information about its past economy.

Trade with the United States has helped northern region cities like Monterrey and Tijuana grow. Foreign-owned factories, called **maquiladoras**, have been built in this region. Many Mexicans cross the border to shop, work, or live in the United States. Some cross the border legally. The U.S. government tries to prevent illegal immigration.

> What has helped Monterrey and Tijuana grow?
>
> _______________________________
>
> _______________________________

Many people in the southern Mexico region speak Indian languages and follow traditional customs. Sugarcane and coffee, two major export crops, grow well in the humid southern climate. Oil production in the region has brought population growth to southern Mexico. Maya ruins, sunny beaches, and clear blue waters make tourism a major industry in the Yucatán Peninsula. Many of today's cities were tiny villages just 20 years ago.

> Underline information about the people in Mexico's southern region.

## CHALLENGE ACTIVITY

**Critical Thinking: Compare and Contrast** Make a four-columned chart—one column for each cultural region in Mexico. Make three rows and write information for each region about History, Population and Economy, and Geography and Natural Resources. When you are done, circle things that are similar among the regions.



 Interactive Reader and Study Guide

# Central America and the Caribbean

## CHAPTER SUMMARY

|  | Central America | Caribbean |
|---|---|---|
| **Geography and Climate** | Isthmus; volcanic, mountainous terrain; tropical | Islands; volcanic or coral reef in origin; tropical |
| **Resources** | Tourism; agriculture |  |
| **History** | Colonized mainly by Spain; all countries became independent | Colonized by various European countries; some islands never became independent |
| **People** | Mainly European or Indian descent, some Africans | European or African descent, some Asians |
| **Languages** | Spanish, native Indian languages, English | Spanish, French, English, Creole |

## COMPREHENSION AND CRITICAL THINKING

Use information from the graphic organizer to answer the following questions.

1. **Identify** What is the main resource in many Caribbean islands? Write your answer in the appropriate row in the graphic organizer.

2. **Contrast** How is the history of Central America different from that of the Caribbean islands?

   _______________________________________________________________

   _______________________________________________________________

3. **Drawing Inferences** Why do you think the native Indian population group is not one of the main groups in the Caribbean islands?

   _______________________________________________________________

   _______________________________________________________________

   _______________________________________________________________

# Central America and the Caribbean

**Section 1**

**MAIN IDEAS**

1. Physical features of the region include volcanic highlands and coastal plains.
2. The climate and vegetation of the region include forested highlands, tropical forests, and humid lowlands.
3. Key natural resources in the region include rich soils for agriculture, a few minerals, and beautiful beaches.

## Key Terms and Places

**isthmus**  narrow strip of land that connects two larger land areas

**Caribbean Sea**  sea surrounded by Central America, the Greater and Lesser Antilles, and South America

**archipelago**  large group of islands

**Greater Antilles**  group of large islands in the Caribbean Sea

**Lesser Antilles**  group of small islands in the Caribbean Sea

**cloud forest**  moist, high-elevation tropical forest where low clouds are common

## Section Summary

### PHYSICAL FEATURES

Central America is an **isthmus** that connects North and South America. It is made up of seven small countries: Belize, Guatemala, El Salvador, Honduras, Nicaragua, Costa Rica, and Panama. At its widest, the isthmus separates the Pacific Ocean and the **Caribbean Sea** by 125 miles (200 km). A chain of mountains and volcanoes runs through the middle of the isthmus. On both sides, a few short rivers run through the coastal plains to the sea. The lack of good water routes and ruggedness of the land make travel difficult.

The Caribbean islands separate the Atlantic Ocean from the Caribbean Sea. On the east lie the **Lesser Antilles**, an **archipelago** of islands that stretch from the Virgin Islands to Trinidad and Tobago. West and north of these are the **Greater Antilles**, which include Cuba, Jamaica, Puerto Rico,

> Underline the seven countries that make up Central America.

> What two bodies of water are separated by the Caribbean islands?
>
> _______________________
>
> _______________________

and Hispaniola. Many of these islands are actually the tops of underwater volcanoes. They are located along the edges of tectonic plates that move against each other, causing earthquakes and volcanic eruptions. The Bahama Islands, located in the Atlantic Ocean, southeast of Florida, were formed by coral reefs.

**What causes earthquakes and volcanoes in the region?**

_______________________________

_______________________________

## CLIMATE AND VEGETATION

Most of the region is generally sunny and warm. Most of Central America's Pacific coast, where plantations and ranches are found, has a tropical savanna climate. The Caribbean coast has areas of tropical rain forest. The inland mountains are cool and humid. Some mountainous areas have dense **cloud forests**, or moist, high-elevation tropical forests where low clouds are common. Many animal and plant species live there.

**Where are cloud forests found?**

_______________________________

_______________________________

Temperatures in the region do not change much from day to night or from winter to summer. Change in seasons is marked by changes in rainfall. Winters are generally dry, but it rains nearly every day in the summer. From summer to fall, hurricanes bring heavy rains and wind, which occasionally cause flooding and great destruction.

## RESOURCES

The region's best resources are its land and climate, which make tourism an important industry. Warm climate and rich volcanic soil make the region a good place to grow coffee, bananas, sugarcane, and cotton. However, the region has few mineral or energy resources.

**What two factors make the region a good place to grow crops?**

_______________________________

_______________________________

## CHALLENGE ACTIVITY

**Critical Thinking: Comparing and Contrasting** Write a description of the year-round climate in your region and compare and contrast it with that of the Central American and Caribbean region.

# Central America and the Caribbean

## Section 2

**MAIN IDEAS**

1. The history of Central America was mostly influenced by Spain.
2. The culture of Central America is a mixture of Native American and European traditions.
3. The countries of Central America today have challenges and opportunities.

## Key Terms and Places

**ecotourism** the practice of using an area's natural environment to attract tourists

**civil war** a conflict between two or more groups within a country

**Panama Canal** a waterway connecting the Pacific Ocean, the Caribbean Sea, and the Atlantic Ocean

## Section Summary

### HISTORY

The Maya people built a civilization in the region from about AD 250 to 900. Many of their descendents, and some of their customs, can still be found. In the 1500s Spain controlled the entire region except for Belize, which became a British colony. The Europeans established gold mines and large tobacco and sugarcane plantations, and forced the Central American Indians to do the hard work. They also brought enslaved Africans to work.

> Underline the country that controlled most of the region after the 1500s.

In 1821 Honduras, El Salvador, Costa Rica, Guatemala, and Nicaragua gained independence and were a single country until 1839. Panama was part of Colombia until 1903. Belize separated from Britain in 1981. Independence did not help most people, as wealthy landowners took control. In the early to mid-1900s the U.S.-based United Fruit Company controlled most of the banana production in the region. Many people resented the role of foreign companies, and thought it was unfair for a few people to have so much power. In the mid- to late 1900s people fought for land reform in Guatemala, El Salvador and Nicaragua.

> Underline the sentences that explain why people in the region objected to large foreign companies.

## CULTURE

Most people in Central America are mestizos, people of mixed Indian and European ancestry. Some Indian people live in areas such as the highlands, and people of African ancestry live mainly along the eastern coast. English is the official language in Belize. Spanish is also spoken there and in the other countries. Many people speak Indian languages.

The Spanish brought Roman Catholicism to people in the region, but Indian traditions are also followed. Corn, tomatoes, chocolate, and hot peppers are foods of the region.

> Underline the languages that are spoken in the region.

## CENTRAL AMERICA TODAY

Guatemala has the region's largest population. Most people are mestizos, but many are descendents of the Maya. Conflict there killed some 200,000 people from 1960 to 1996. Coffee is the most important crop.

Belize has the region's lowest population. Recently, **ecotourism** has become a large industry, as more people visit the Maya ruins and coral reefs.

Honduras is mountainous, making transportation difficult. It has little farmland, but fruit is exported.

In the 1980s the poor people of El Salvador fought a **civil war** against the few rich families that owned much of the best land, which is very fertile. The war ended in 1992, and people are rebuilding.

Nicaragua was ruled by Sandinistas from 1979 to 1990. After a civil war, it became a democracy.

Costa Rica has been more peaceful than most of its neighbors, which helps its economy. Coffee, bananas, and tourism are its largest industries.

Most Panamanians live near the **Panama Canal**, built and controlled by the United States. In 1999 the U.S. gave control of the canal to Panama.

> Underline the countries that fought a civil war in the last century.

> Which country in the region has been the most stable?
> _______________________

## CHALLENGE ACTIVITY

**Critical Thinking: Sequence** Make a time line that shows important events in Central American history.

# Central America and the Caribbean

Section 3

**MAIN IDEAS**

1. The history of the Caribbean islands includes European colonization followed by independence.

2. The culture of the Caribbean islands shows signs of past colonialism and slavery.

3. Today the Caribbean islands have distinctive governments with economies that depend on agriculture and tourism.

## Key Terms and Places

**dialect** regional variety of a language

**commonwealth** self-governing territory associated with another country

**refugee** someone who flees to another country, usually for political or economic reasons

**Havana** the capital of Cuba

**cooperative** organization owned by its members and operated for their benefit

## Section Summary

### HISTORY

The Caribbean islands were the first land Christopher Columbus saw in 1492, though he thought he had sailed to islands near India. By the 1700s the islands were colonized by the Spanish, English, French, Dutch, and Danish. They brought enslaved Africans to work on their sugarcane plantations.

> Underline the sentence that tells who colonized the Caribbean Islands.

Led by Toussaint L'Ouverture, Haiti gained independence in 1804. Cuba became independent in 1902. Although other islands became independent after World War II, some, such as Martinique and Guadeloupe, never did. Most people on these French islands do not want independence.

### CULTURE

Most islanders are descended from Europeans, Africans, or a mixture of the two. There are also some Asians. Some people speak English, French or Spanish—others speak Creole, a **dialect**, or regional variety of a language.

Today, islands colonized by France and Spain have many Catholics. In Cuba and elsewhere,



 Interactive Reader and Study Guide

**Section 3, *continued***

people practice a blend of Catholicism and traditional African religion called Santería. Caribbeans enjoy the Carnival holiday. It comes before the Christian season of Lent and features parades and costumes. Some of the region's foods, such as okra and yams, were brought by enslaved Africans. They also made souse, a dish made from the leftover pig parts given to them by slaveholders.

## THE CARIBBEAN ISLANDS TODAY

Puerto Rico is a **commonwealth**, a self-governing territory associated with the United States. Some Puerto Ricans are happy about this, while others would like to become a state or a separate country. Though richer than others in the region, Puerto Ricans are not as well off as other U.S. citizens.

Haiti, occupying the western part of Hispaniola, is the poorest country in the Americas. Dishonest governments have caused violence, which many Haitians have tried to escape by becoming **refugees**.

The eastern part of Hispaniola is the Dominican Republic. Its industries are tourism and agriculture.

**Havana** is the capital of Cuba, the largest, most populous island in the Caribbean. Fidel Castro has been the leader of Cuba's Communist government since 1959. In Cuba, the government runs the economy, newspapers and television. Farms are organized as **cooperatives**, owned by the people who work on them and run for their benefit.

Some of the rest of the Caribbean islands, such as Jamaica, are independent countries, while others, such as the Virgin Islands, are territories of other countries. Most rely on tourism for their income.

## CHALLENGE ACTIVITY

**Critical Thinking: Drawing Inferences** Imagine you are a Haitian refugee. Write a paragraph describing the reasons you left Haiti and your hopes for the future.

> Underline the sentence that describes the general economic status of people living in Puerto Rico.

> What is the poorest country in the Americas?
> _______________________

> Underline the sentences that describe Cuba's government.

# Caribbean South America

## CHAPTER SUMMARY

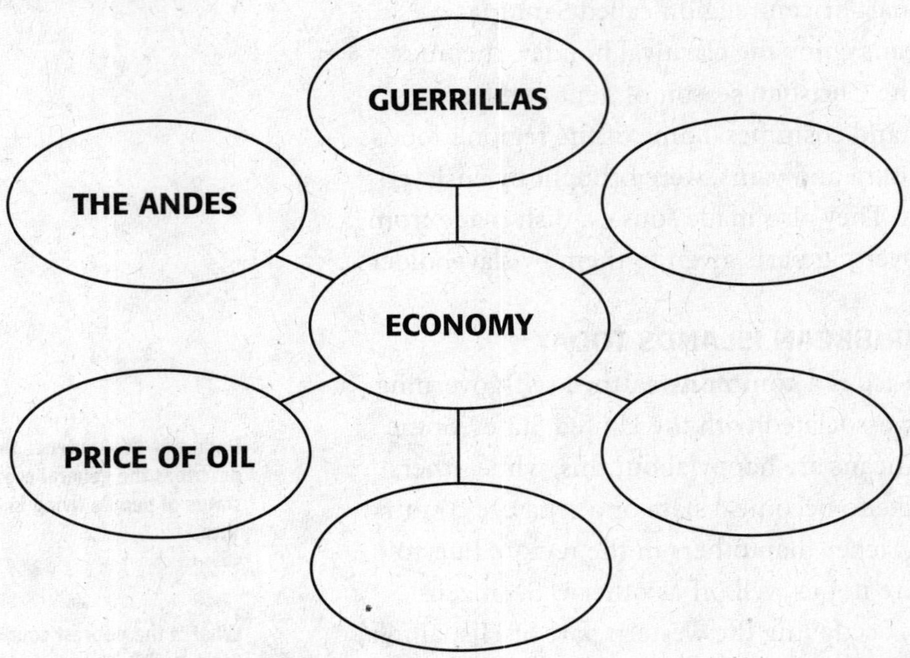


## COMPREHENSION AND CRITICAL THINKING

Use information from the graphic organizer to answer the following questions.

**1. Identify** Which of the three factors affecting the economy of Caribbean South America shown in the graphic organizer has the most impact?

_______________________________________________

_______________________________________________

**2. Interpret** Choose three of the following terms that have the most impact on the region's economy and write them in the graphic organizer: *piranhas, Caracas slums, Cauca River, strike, El Dorado, Gran Colombia, indigo, joropo, coffee.*

**3. Make Inferences** Name a factor that will never stop having an effect on the region's economy.

_______________________________________________

_______________________________________________

_______________________________________________

# Caribbean South America

### Section 1

**MAIN IDEAS**

1. Caribbean South America has a wide variety of physical features and wildlife.
2. The region's location and elevation both affect its climate and vegetation.
3. Caribbean South America is rich in resources, such as farmland, oil, timber, and rivers for hydroelectric power.

## Key Terms and Places

**Andes**  mountains on the western side of Colombia

**cordillera**  a mountain system made up of roughly parallel ranges

**Guiana Highlands**  plateaus stretching from Venezuela to Suriname

**Llanos**  plains region between the highlands and Andes

**Orinoco River**  longest river in the region, flows through Venezuela to the Atlantic Ocean

## Section Summary

### PHYSICAL FEATURES AND WILDLIFE

Caribbean South America includes rugged mountains, highlands, and plains drained by huge river systems. The region's highest point is in western Colombia, where the snow-capped **Andes** reach 18,000 feet. These mountains form a **cordillera**, or a system of roughly parallel mountain ranges. Some peaks in the Andes are active volcanoes. The **Guiana Highlands** are a series of plateaus stretching from Venezuela to Suriname. Some areas have flat-topped layers called *tepuís*, made of sandstone that has resisted erosion.

> **Where are the Guiana Highlands?**
> _______________________
> _______________________

The **Llanos** is a region of plains that lies between the Andes and the Guiana Highlands. The Llanos is mostly grassland and often floods. The **Orinoco River**, the region's longest, flows for about 1600 miles and empties into the Atlantic Ocean in eastern Venezuela. The Orinoco and its tributaries drain the plains and highlands. Two other rivers, the Cauca and the Magdalena, drain the Andean region.

> **Underline the sentence that explains what the Llanos is.**

> **What two rivers drain the Andean region?**
> _______________________
> _______________________

## Section 1, *continued*

Wildlife in Caribbean South America includes meat-eating fish called piranhas, jaguars, ocelots, monkeys, and hundreds of bird species.

## CLIMATE AND VEGETATION

Most of Caribbean South America is warm year-round due to its location near the equator. The coldest temperatures are found at high elevations in the Andes. Temperatures can fall as much as four degrees for every additional 1,000 feet above sea level. The grassy Llanos has low elevation, giving it a tropical savanna climate with wet and dry seasons. Humid tropical rain forests cover much of southern Colombia. Heavy rainfall produces huge trees and vegetation so thick that sunlight barely penetrates to the jungle floor.

> **How much can temperature change at higher elevations?**
> ___________________________
> ___________________________

## RESOURCES

Resources in Caribbean South America include rich soils, oil reserves in both Venezuela and Colombia, timber from the forests, fish and shrimp from coastal waters, and plentiful supplies of hydroelectric power.

> **Circle the resources of Caribbean South America that are mentioned in the paragraph.**

## CHALLENGE ACTIVITY

**Critical Thinking: Evaluating** Write a brief report that identifies the factors that affect the climate of Caribbean South America and explains how the climate changes throughout the region.

# Caribbean South America

## Section 2

**MAIN IDEAS**

1. Native cultures, Spanish conquest, and independence shaped Colombia's history.
2. In Colombia today, the benefits of a rich culture and many natural resources contrast with the effects of a long period of civil war.

## Key Terms and Places

**Cartagena** major Caribbean naval base and port during the Spanish empire

**Bogotá** capital of Colombia, located high in the eastern Andes

**guerrillas** members of an irregular military force

## Section Summary

### COLOMBIA'S HISTORY

Before the Spanish conquest around 1500, the Chibcha people lived in Colombia. The Chibcha had a well-developed civilization. They made pottery, wove fabrics, and made fine objects from gold and other metals. To expand their new empire, the Spanish defeated the Chibcha and seized much of their treasure. The Spanish claimed the land, started a colony, and built large estates, forcing the Chibcha and enslaved Africans to work on them. They also built a large fort and commercial port at the city of **Cartagena**.

Who was forced to work for the Spanish in Colombia?

_______________________________

_______________________________

By the late 1700s many people in the Spanish colonies began struggling for independence. The people who lived in the land that is now Ecuador, Panama, Venezuela, and Colombia formed a Republic called Gran Colombia. The republic broke up, and in 1830 New Granada, which included Colombia and Panama, was created. Even after independence there was trouble in Colombia. People could not agree about how much power the Catholic Church or the central government should have. Violence in Colombia broke out from time to time, and continues to the present day. Part of the

Underline the names of the countries that were part of Gran Colombia.



    Interactive Reader and Study Guide

## Section 2, *continued*

problem is Colombia's rugged geography, which isolates people into separate regions. Many people identify with their region more than their nation.

## COLOMBIA TODAY

Most Colombians live in the fertile valleys and river basins among the mountain ranges, where the climate is moderate. **Bogotá**, the capital, is located high in the eastern Andes. Few people live in the rain forests of the south. Each region has a distinct geography as well as a distinct culture. Most people are Roman Catholic and speak Spanish. Dances and songs reflect the cultures of each region. Culture on the Caribbean coast is influenced by African culture while South American Indian culture survives in remote mountain areas. Most people are mestizos or of Spanish, African, or Indian descent.

Colombia's major export is oil. It also supplies most of the world's emeralds. Other minerals include iron ore, coal, and gold. Colombia's most famous agricultural product is coffee, but it also grows bananas, sugarcane, corn, rice, and flowers.

Today, different groups fight each other and the government. One army of **guerrillas**, an irregular military force, wants to overthrow the government. Guerrillas have seized land from farmers and are involved in growing the illegal coca plant, which is used to make the dangerous drug cocaine. Colombia's government continues to fight the guerrillas with laws and military action. The United States government provides money and equipment to help in this effort.

## CHALLENGE ACTIVITY

**Critical Thinking: Drawing Inferences** On a separate piece of paper, write a paragraph describing how life in Colombia might be different if there were no mountains there.

## Caribbean South America

### Section 3

**MAIN IDEAS**

1. Spanish settlement shaped the history and culture of Venezuela.
2. Oil production plays a large role in Venezuela's economy and government today.
3. The Guianas have diverse cultures and plentiful resources.

## Key Terms and Places

**llaneros** Venezuelan cowboys

**Lake Maracaibo** a lake near the Caribbean Sea, rich in oil deposits

**Caracas** the capital of Venezuela and the economic and cultural center of the country

**strike** a group of workers stopping work until their demands are met

**referundum** a recall vote

## Section Summary

### HISTORY AND CULTURE OF VENEZUELA

The Spanish came to Venezuela in the early 1500s
looking for gold, but found little. So they grew
indigo, a plant used to make blue dye. They forced
Indians and enslaved Africans to do the hard work.
In the early 1800s the Venezuelan people, led by
Simon Bolívar, revolted against their Spanish rulers
and fought for independence, which they officially
gained in 1830. Military dictators ran the country
throughout the 1800s. Oil was discovered in the
1900s, but brought wealth only to the powerful.

> When did Venezuela officially become independent?
>
> _______________________

Venezuelans are of native Indian, African, and
European descent. European descendents tend to
live in the cities, and African descendents tend to
live on the coast. Most people are Spanish-speaking
Roman Catholics, but Indian languages and religious
beliefs have been kept alive. Venezuelans enjoy soccer,
baseball, rodeos, and the national dance, the joropo.

> Underline the different groups who live in Venezuela.

### VENEZUELA TODAY

Many people in rural Venezuela are farmers or
ranchers. The ranchers are called **llaneros**—cowboys
of the Llanos. Venezuela's economy is based on oil,



 Interactive Reader and Study Guide

## Section 3, *continued*

found mostly near the Orinoco River and **Lake Maracaibo**. The oil industry has made some people wealthy and attracted many immigrants; however, the vast majority of the population lives in poverty. **Caracas**, Venezuela's capital, is the country's economic and cultural center. It has modern subways and buildings, but it is surrounded by slums.

Military dictators ran Venezuela until 1959, when the first president was elected. Since then the government has dealt with economic and political problems. In 2002 President Hugo Chavez started to distribute the country's oil income equally among all Venezuelans. Millions of Venezuelans went on **strike** for two months. The country's economy suffered greatly. In 2004 Venezuelans opposed to President Chavez called for a **referendum**, a vote that would remove him from office, but it failed. Recently Chavez created new policies to end poverty, illiteracy, and hunger in Venezuela.

> **Why do you think there is significant poverty in Venezuela?**
> ___________________________

> **What was the result of Venezuela's referendum in 2004?**
> ___________________________

## THE GUIANAS

The Guianas consist of the countries Guyana, Suriname, and French Guiana. These countries have diverse populations with many people descended from Africans, but each country is different. Guyana has many immigrants from India who came to work on sugarcane plantations. Today most people run small farms or businesses. Suriname's population includes Creoles, or people of mixed heritage, and people from China, Indonesia, Africa, and South Asia. Suriname's economy is similar to Guyana's. French Guiana is a territory of France. Most people live in coastal areas, and the country relies heavily on imports.

> **Circle names of the countries that make up the Guianas.**

**Critical Thinking: Identifying Points of View** Why would some Venezuelans think Hugo Chavez should no longer be president? Write a short paragraph to explain your answer.

# Atlantic South America

## CHAPTER SUMMARY

**Atlantic South America**

| | Brazil | Argentina | Uruguay | Paraguay |
|---|---|---|---|---|
| **History** | | | | |
| **People & Culture** | | | | |
| **Resources** | | | | |
| **Economy** | | | | |

## COMPREHENSION AND CRITICAL THINKING

Use the graphic organizer to answer the following questions.

**1. Identify** In the categories above, where would you write notes about how Brazil gained independence? Where would you write facts about the main religion practiced in Paraguay?

_______________________________________________

_______________________________________________

**2. Interpret** Which two of the following words do not belong in the "Economy" category: *informal economy, Mercosur, economic crisis, gaucho, landlocked*?

_______________________________________________

_______________________________________________

**3. Sequence** List the Atlantic South American countries in order by size—largest to smallest.

_______________________________________________

_______________________________________________

# Atlantic South America

**Section 1**

**MAIN IDEAS**

1. Physical features of Atlantic South America include large rivers, plateaus, and plains.
2. Climate and vegetation in the region ranges from cool, dry plains to warm, humid forests.
3. The rain forest is a major source of natural resources.

## Key Terms and Places

**Amazon River**  4,000 mile-long river that flows eastward across northern Brazil

**Río de la Plata**  an estuary that connects the Paraná River and the Atlantic Ocean

**estuary**  a partially enclosed body of water where freshwater mixes with salty seawater

**Pampas**  wide, grassy plains in central Argentina

**deforestation**  the clearing of trees

**soil exhaustion**  soil that has become infertile because it has lost nutrients needed by plants

## Section Summary

### PHYSICAL FEATURES

The region of Atlantic South America includes four countries: Brazil, Argentina, Uruguay, and Paraguay. A major river system in the region is the Amazon. The **Amazon River** extends from the Andes Mountains in Peru to the Atlantic Ocean. The Amazon carries more water than any other river in the world.

The Paraná River, which drains much of the central part of South America, flows into an **estuary** called the **Río de la Plata** and the Atlantic Ocean.

The region's landforms mainly consist of plains and plateaus. The Amazon Basin in northern Brazil is a huge, flat floodplain. Further south are the Brazilian Highlands and an area of high plains called the Mato Grosso Plateau.

The low plains region, Gran Chaco, stretches across parts of Paraguay and northern Argentina. The grassy plains of the **Pampas** are found in

> **What four countries make up Atlantic South America?**
> ___________________________
> ___________________________

> **What is the Amazon Basin?**
> ___________________________
> ___________________________

> **Circle the names of three plains regions in Atlantic South America.**



  Interactive Reader and Study Guide

**Section 1, *continued***

central Argentina. Patagonia is a region of
dry plains and plateaus south of the Pampas.
These plains rise in the west to form the Andes
Mountains.

## CLIMATE AND VEGETATION

Atlantic South America has many climates.
Southern and highland areas have cool climates
while northern and coastal areas have tropical and
moist climates.

Patagonia has a cool, desert climate. To the north
in the Pampas, the climate is humid subtropical. In
Argentina, the Gran Chaco has a humid tropical
climate. Central Brazil has a tropical savanna
climate, but to the northeast the climate is hot and
dry. In southeast Brazil, the climate is cooler and
more humid. In northern Brazil, where the Amazon
Basin is located, the climate is humid tropical.

**What type of climate does northern Brazil have?**

______________________________

______________________________

## NATURAL RESOURCES

The Amazon rain forest is one of Atlantic South
America's greatest natural resources. It provides
food, wood, rubber, plants for medicines, and other
products. **Deforestation** has become an issue that
threatens the resources of the rain forest.

**What threatens the Amazon rain forest?**

______________________________

Land near the coastal areas in the region is
used for commercial farming. However, planting
the same crop year after year has caused **soil
exhaustion** in some areas.

Other resources in Atlantic South America
include gold, silver, copper, iron, and oil. Dams on
some rivers also provide hydroelectric power.

## CHALLENGE ACTIVITY

**Critical Thinking: Drawing Inferences** Write an essay describing ways to
preserve the Amazon rain forest while still helping people who rely on
its resources to survive.



  Interactive Reader and Study Guide

# Atlantic South America

## Section 2

**MAIN IDEAS**

1. Brazil's history has been affected by Brazilian Indians, Portuguese settlers, and enslaved Africans.
2. Brazil's society reflects a mix of people and cultures.
3. Brazil today is experiencing population growth in its cities and new development in rain forest areas.

## Key Terms and Places

**São Paulo**  the largest urban area in South America, located in southeastern Brazil

**megacity**  a giant urban area that includes surrounding cities and suburbs

**Rio de Janeiro**  Brazil's second-largest city, located northeast of São Paulo

**favelas**  huge slums

**Brasília**  the capital of Brazil

**Manaus**  a major port and industrial city; located 1,000 miles from the mouth of the Amazon River

## Section Summary

### HISTORY

Brazil is the largest country in South America. It has a population of more than 186 million people. The first people in Brazil were American Indians who arrived in the region thousands of years ago.

Who were the first people to live in Brazil?

_______________________

_______________________

In 1500 Portuguese explorers became the first Europeans to find Brazil. Colonists brought Africans to the region to work as slaves on sugar plantations. These plantations helped make Portugal rich.

Gold and precious gems were discovered in the late 1600s and early 1700s in the southeast, and a mining boom drew people to Brazil from all over the world. Brazil became a major coffee producer in the late 1800s. Brazil gained independence from Portugal without a fight in 1822. Since the end of Portuguese rule Brazil has been governed by both dictators and elected officials. Today Brazil has an elected president and legislature.

What discovery brought people to Brazil from all over the world?

_______________________

_______________________



     Interactive Reader and Study Guide

## PEOPLE AND CULTURE

Nearly 40 percent of Brazil's people are of mixed African and European descent. Brazil also has the largest Japanese population outside of Japan. Brazil's official language is Portuguese.

Brazil has the world's largest population of Roman Catholics. Some Brazilians also practice macumba, a religion that combines beliefs and practices of African and Indian religions with Christianity.

Brazilians celebrate Carnival, a celebration that mixes traditions from Africa, Brazil, and Europe. Immigrant influences can also be found in Brazilian foods.

> **What percentage of Brazil's people are of mixed African and European descent?**
> ________________________

> **What is macumba?**
> ________________________
> ________________________

## BRAZIL TODAY

Brazil can be divided into four regions. The southeast is the most populated. More than 17 million people live in the city of **São Paulo**, which is considered a **megacity**, or giant urban area that includes surrounding cities and suburbs. **Rio de Janeiro** is Brazil's second-largest city, also located in the southeast. The southeast has a good economy, but it also has poverty. Many people live in city slums, or **favelas**.

> **What is Brazil's second-largest city?**
> ________________________

The northeast is Brazil's poorest region. Drought has made farming difficult. However, many tourists are attracted to the region.

The interior region is a frontier land, with much potential for farming. The capital of Brasil, **Brasília**, is located there.

The Amazon region covers the northern part of Brazil. **Manaus** is a major port and industrial city 1,000 miles from the mouth of the Amazon River. The Amazon rain forest is a valuable resource to people who live and work there, but deforestation threatens the wildlife and Brazilian Indians living there.

> **Circle the word describing an important industry in northeast Brazil.**

> **What are the four regions of Brazil?**
> ________________________
> ________________________

## CHALLENGE ACTIVITY

**Critical Thinking: Drawing Inferences** Why do most Brazilians live in the southeast? Turn to a partner and give reasons for your answer.

# Atlantic South America

**Section 3**

**MAIN IDEAS**

1. European immigrants have dominated the history and culture of Argentina.
2. Argentina's capital, Buenos Aires, plays a large role in the country's government and economy today.
3. Uruguay has been influenced by its neighbors.
4. Paraguay is the most rural country in the region.

## Key Terms and Places

**gauchos** Argentine cowboys

**Buenos Aires** the capital of Argentina

**Mercosur** an organization that promotes trade and economic cooperation among the southern and eastern countries of South America

**informal economy** a part of the economy based on odd jobs that people perform without government regulation through taxes

**landlocked** completely surrounded by land with no direct access to the ocean

## Section Summary

### ARGENTINA'S HISTORY AND CULTURE

Argentina was originally home to groups of Indians. In the 1500s, Spanish conquerors spread into southern South America in search of silver and gold. They built settlements in Argentina. Spanish monarchs granted land to the colonists. Landowners forced Indians living there to work.

During the colonial era, the Pampas became an important agricultural region. Argentine cowboys called **gauchos** herded cattle and horses there.

Argentina fought and gained independence in the 1800s. Many Indians were killed. Immigrants from Italy, Germany, and Spain began to arrive in Argentina.

During the "Dirty War" in the 1970s, many Argentines were tortured and killed after being accused of disagreeing with the government. Argentina's military government gave up power to an elected government in the 1980s.

> **Why did Spanish conquerors come to the region in the 1500s?**
>
> _______________________
>
> _______________________

> **Immigrants began to arrive from what European countries?**
>
> _______________________
>
> _______________________



   Interactive Reader and Study Guide

## ARGENTINA TODAY

**Buenos Aires** is the capital of Argentina. It is the second-largest urban area in South America. In the 1990s, government leaders made economic reforms to help businessses grow. Argentina joined **Mercosur**—an organization that promotes trade and economic cooperation among the southern and eastern countries of South America. However, heavy debt and government spending brought Argentina into an economic crisis. Many people lost their jobs and joined the **informal economy**—a part of the economy based on odd jobs that people perform without government regulation through taxes.

> **What is the purpose of Mercosur?**
> ________________________
> ________________________
> ________________________

## URUGUAY

Uruguay lies between Argentina and Brazil. Its capital is Montevideo. Portugal claimed Uruguay during the colonial era, but the Spanish took over by the 1770s. Few Uruguayan Indians remained. Uruguay declared its independence from Spain in 1825. Today Uruguay is a democracy.

> **What is the capital of Uruguay?**
> ________________________

## PARAGUAY

Paraguay is **landlocked**, which means completely surrounded by land, with no direct access to the ocean. It was claimed by the Spanish in the mid-1530s and remained a Spanish colony until 1811, when it won indepedence. Today Paraguay is a democracy. Ninety-five percent of Paraguayans are mestizos. People of European descent and Indians make up the rest of the population. Most Paraguayans speak both Spanish and Guarani. Agriculture is an important part of the economy.

> **Underline the definition of** *landlocked*.

> **What type of government do Uruguay and Paraguay have today?**
> ________________________

## CHALLENGE ACTIVITY

**Critical Thinking: Drawing Inferences** Which Atlantic South American country do you think has the strongest economy? Write a paragraph giving reasons to support your answer.

Name _________________________________ Class _______________ Date ______________

# Pacific South America

## CHAPTER SUMMARY

| Countries of Pacific South America | |
| --- | --- |
| **Similarities** | **Differences** |
| **1.** Andes | **1.** The Atacama Desert is in northern ___________________________. |
| **2.** Inca empire | **2.** ___________________ has large oil and gas reserves. |
| **3.** Mix of Spanish and indigenous people | **3.** ___________________ has the highest percentage of South American Indians. |
| **4.** Political instability | **4.** ___________________ is the largest and most populous country. |
| **5.** ___________________ | **5.** ___________________ is the poorest country. |

## COMPREHENSION AND CRITICAL THINKING

Use information from the graphic organizer to answer the following
questions.

1. **Identify** What economic problem do all the countries have? Write your answer in the appropriate space on the graphic organizer.

2. **Identify** Complete the sentences in the "Differences" column with the name of the correct country in Pacific South America.

3. **Analyze** Which items in both columns have to do with physical geography?

_______________________________________________________________

_______________________________________________________________

_______________________________________________________________



 Interactive Reader and Study Guide

# Pacific South America

## Section 1

**MAIN IDEAS**

1. The Andes Mountains are the main physical feature of Pacific South America.
2. The region's climate and vegetation change with elevation.
3. Key natural resources in the region include lumber, oil, and minerals.

## Key Terms and Places

**altiplano**  a broad, high plateau that lies between the ridges of the Andes

**strait**  a narrow body of water that connects two larger bodies of water

**Atacama Desert**  a very dry desert in northern Chile

**El Niño**  an ocean weather pattern that affects the Pacific coast

## Section Summary

### PHYSICAL FEATURES

The Andes Mountains run through all of the Pacific
South American countries, with peaks above 20,000
feet. In the south, the mountains are rugged and
covered by ice caps. In the north, they are rounded,
and the range splits into two ranges. A high plateau,
called the **altiplano**, lies between the two ranges.
Earthquakes and volcanoes sometimes disturb
Andean glaciers, causing ice and mud slides.

> Underline the differences between the Andes Mountains in the south and in the north.

The region has Amazon tributaries, but few other
major rivers. Rivers flowing into the altiplano never
reach the sea, but fill two large lakes. Between the
southern tip of Chile and Tierra del Fuego lies the
Strait of Magellan. A **strait** is a waterway linking
two large bodies of water. Ecuador's Galápagos
Islands in the Pacific Ocean have wildlife not found
anywhere else in the world.

### CLIMATE AND VEGETATION

Climate varies more with elevation than with
latitude in this region. There are five climate zones,
starting at the hot and humid zone near sea level,
where bananas and sugarcane are grown. Coffee is
the main crop of the second climate zone, with

> What is the main reason climate varies in the region?
>
> _______________________
>
> _______________________



   Interactive Reader and Study Guide

**Section 1,** *continued*

cooler, moist air and mountain forests. Many large cities are located there. Many people also live in the third, cooler climate zone. Some grow potatoes and wheat among the grasslands and forests. Grassland and shrubs, but no trees, grow in the fourth zone. In the highest zone, near the mountaintops, no vegetation grows, and most of this zone is covered in snow year-round.

> Underline details that describe the five climate zones in the region.

The **Atacama Desert** in northern Chile receives rain only about once every 20 years. Fog and low clouds blow in from the Pacific Ocean, making this coastal desert one of the driest, cloudiest places on Earth. In Peru, snow melting in the Andes collects in rivers flowing down to the Pacific Ocean, allowing a small number of people to live there.

> How often does it rain in the Atacama Desert?
>
> _______________________________

Every two to seven years, the water off the Pacific coast is warm, rather than cool. This change of pattern, called **El Niño**, causes the fish off the coast to find cooler waters, leaving local fishermen with less to catch. El Niño also causes extreme and unusual weather around the world.

## NATURAL RESOURCES

The region's natural resources include lumber from forests in southern Chile, eastern Peru, and Ecuador. Bolivia produces the minerals tin, gold, silver, lead, and zinc. Chile exports more copper than any other country in the world. Ecuador's main export is oil, and it also has reserves of natural gas. Most of the region, however, has little good farmland, and so there are few agricultural exports.

> Underline minerals found in Bolivia.

## CHALLENGE ACTIVITY

**Critical Thinking: Comparing and Contrasting** Create an illustrated brochure that tells about the five climate zones in the region.

# Pacific South America

Section 2

---

**MAIN IDEAS**

1. The countries of Pacific South America share a history influenced by the Inca civilization and Spanish colonization.
2. The culture of Pacific South America includes American Indian and Spanish influences.

## Key Terms

**viceroy** governor of a Spanish colony
**Creoles** American-born descendants of Europeans

## Section Summary

### HISTORY

Peru's first advanced people lived in the Andes about 900 BC. They raised crops in stone terraces built into the steep mountainsides. Near the coast, other people used irrigation to control flooding and store water. These early people supported themselves and their towns by growing crops. One early mountain culture, the Tiahuanaco, made huge stone carvings near a lake. Other people, near the Peruvian coast, scratched the outlines of animals and other shapes into the surface of the desert. These designs, called the Nazca Lines, are so large that they can only be seen from high above.

> Underline the sentence about the farming methods of Peru's first advanced people.

By the early 1500s, the Incas controlled the area from northern Ecuador to central Chile. The Inca Empire was the home of about 12 million people. They built stone-paved roads and suspension bridges across the steep Andean valleys. They used irrigation to turn the desert into farmland. Although the Incas were an advanced civilization, they had no written language, no wheeled vehicles, and no horses. Teams of runners carried messages, as in a relay race, to distant parts of the empire. One team could cover as much as 150 miles a day.

> How many people were in the Inca Empire by the early 1500s?
> ____________________

One day in the early 1500s, a new Inca ruler was on his way to be crowned. He met the Spanish





Interactive Reader and Study Guide

**Section 2,** *continued*

explorer Francisco Pizarro, who took him captive. The new Inca ruler ordered a room to be filled with gold and silver objects, thinking it was to be his ransom. Instead, Pizarro had the Inca ruler killed. By 1535 the Spanish had conquered the entire Inca Empire. The Spanish forced many Indians to work in mines or on plantations. The Spanish **viceroy**, or governor, made the Indians follow Spanish laws and customs.

> Underline the sentence that tells what the new Inca ruler did after being taken captive by Pizarro.

By the early 1800s people began to revolt against Spanish rule. **Creoles**, American-born descendants of Europeans, were the main leaders. These revolts helped Chile, Ecuador, Bolivia, and Peru gain independence by 1825.

## CULTURE

Most people in the region speak Spanish, an official language in every country in the region. However, millions of South American Indians speak a native language. Some speak only their native language, and some also speak Spanish. Bolivia, which has the highest percentage of Indians in South America, has three official languages: Spanish and two Indian languages.

> Which country has three official languages?
> _______________________

The religion of this region also reflects both Spanish and native Indian influences. Roman Catholic traditions come from the Spanish, but some people in the Andes still practice ancient religious customs. In June, for example, they participate in a sun worship festival that was celebrated by the Incas. They wear traditional costumes, sometimes with wooden masks. They also play wooden flutes and other traditional instruments.

> Underline the sentence that tells about an ancient Inca religious festival.

## CHALLENGE ACTIVITY

**Critical Thinking: Comparing** Write a short essay comparing the Inca civilization before and after the Spanish conquest.

# Pacific South America

**Section 3**

**MAIN IDEAS**

1. Ecuador struggles with poverty and political instability.
2. Bolivia's government is trying to gain stability and improve the economy.
3. Peru has made progress against poverty and violence.
4. Chile has a stable government and a strong economy.

## Key Terms and Places

**Quito**  the capital of Ecuador

**La Paz**  one of Bolivia's two capital cities

**Lima**  the capital of Peru

**coup**  the sudden overthrow of a government by a small group of people

**Santiago**  the capital of Chile

## Section Summary

### ECUADOR TODAY

Although Ecuador is a democracy, it has not had a stable or popular government for a long time. The country has had 9 different presidents in 10 years. In 2005 the president was forced out of office because reforms had not improved conditions for the citizens of Ecuador.

> Underline the sentence that tells why the president of Ecuador was forced out of office in 2005.

Ecuador has three economic regions. Most of the agriculture and industry is found along the coastal plain region. The Andes region, where the capital, **Quito**, is located, is poor. Tourism is a major industry there. The third region, the Amazon basin, produces Ecuador's major export, oil.

### BOLIVIA TODAY

Bolivia has two capitals—the supreme court meets in Sucre and the congress meets in **La Paz**, the highest capital in the world. After many years of military rule, Bolivia is now a democracy. Many people do not think the government is doing enough to help the poor. Several presidents have had to resign because of political protests, and development

> What are Bolivia's two capitals?
> ______________________

## Section 3, *continued*

has been slow. Bolivia is the poorest country in
South America, although it does have resources
such as metals and natural gas.

### PERU TODAY

Peru is the largest and most populous country in
Pacific South America, and its capital, **Lima**, is the
region's largest city. Lima was the colonial capital
of Peru, and has many beautiful buildings. Many
Peruvians have moved there to find jobs in
government or industry. However, many people in
Lima are poor, living in slums around the city center.
Often they cannot get water or electricity. Other
poor people have built houses at the edge of the
city in areas called "young towns." In the 1980s and
1990s, a group called the Shining Path terrorized
people, especially in the highlands. The government
fought it, and Peru is making economic progress.
Mineral deposits are located near the coast, and
hydroelectric projects provide energy. In the
highlands, people grow potatoes and corn.

> Underline the sentences that describe the living conditions of poor people in Lima.

> What group terrorized people living in the highlands in the 1980s and 1990s?
>
> _________________________

### CHILE TODAY

In the 1970s Chile's elected president was
overthrown in a U.S.-backed military **coup**.
These military leaders later imprisoned or killed
thousands of their political opponents. In the late
1980s, Chileans restored democracy, and now they
have one of the most stable governments in South
America. About one third of Chileans live in central
Chile, near the capital, **Santiago**. This area has a
Mediterranean climate good for crops such as grapes
for making wine. Besides farming, fishing, and
forestry, Chile's resources include copper mining,
which accounts for one-third of Chile's exports.

> Underline Chile's resources.

**Critical Thinking: Comparing and Contrasting** You are a recent arrival
to a "young town" in Lima. Write a letter to a friend in another country
about Lima and your home.



                                              Interactive Reader and Study Guide

# The United States

## CHAPTER SUMMARY

| PHYSICAL GEOGRAPHY | HISTORY AND CULTURE | THE UNITED STATES TODAY |
|---|---|---|
| **Physical Features**<br>The East and South<br>The Interior Plains<br>The West | **First Modern Democracy**<br>American Colonies<br>Expansion/Growth<br>Wars and Peace<br>Government/Citizenship<br>Rights/Responsibilities | **Regions**<br>The Northeast<br>The South<br>The Midwest<br>The West |
| **Climate**<br>The East and South<br>The Interior Plains<br>The West | **People and Culture**<br>Ethnic Groups<br>Language<br>Religion<br>Foods and Music<br>Popular Culture | **Economic, Military, and Terrorism**<br>Economic/Military<br>Terrorism |
| **Natural Resources**<br>The East and South<br>The Interior Plains<br>The West | | |

## COMPREHENSION AND CRITICAL THINKING

Use information from the graphic organizer to answer the following
questions.

1. **Interpret** In the categories above, where would you put notes about the climate in
   the West? Where would you put notes about the Revolutionary War?

   _________________________________________________________________________

2. **Identify** Of the following five words (humid, mountainous, plains, subarctic,
   tundra), which three belong under the sub-category "Climate"?

   _________________________________________________________________________

3. **Sequence** If you were going to describe the United States by region, starting in
   the Northeast and working westward, in what order would you describe them?

   _________________________________________________________________________

   _________________________________________________________________________

# The United States

**Section 1**

> ### MAIN IDEAS
> 1. Major physical features of the United States include mountains, rivers, and plains.
> 2. The climate of the United States is wetter in the East and South and drier in the West.
> 3. The United States is rich in natural resources such as farmland, oil, forests, and minerals.

## Key Terms and Places

**Appalachian Mountains**  the main mountain range in the East

**Great Lakes**  the largest group of freshwater lakes in the world

**Mississippi River**  North America's longest river

**tributary**  a stream or river that flows into a larger stream or river

**Rocky Mountains**  an enormous mountain range in the West

**continental divide**  an area of high ground that divides the flow of rivers towards opposite ends of a continent

## Section Summary

### PHYSICAL FEATURES

The United States is one of the largest countries in the world. On the eastern coast of the United States, the land is flat and close to sea level. This area is called the Atlantic Coastal Plain. Moving west, the land rises to a region called the Piedmont. The land rises higher in the **Appalachian Mountains**, the main mountain range in the East. The highest peak in the Appalachians is about 6,700 feet (2,040m).

West of the Appalachian Mountains are the Interior Plains. The plains are filled with rolling hills, lakes, and rivers. The main physical features of the Interior Plains are the **Great Lakes**. The Great Lakes are the largest freshwater lake system in the world. They are also an important waterway for trade between the United States and Canada.

The **Mississippi River** lies west of the Great Lakes. It is the longest river in North America.

**Where is the Atlantic Coastal Plain?**

**What is special about the Great Lakes?**

**What is special about the Mississippi River?**



   Interactive Reader and Study Guide

**Section 1,** *continued*

**Tributaries** of the Mississippi River deposit rich silt that produces fertile farmlands. These farmlands cover most of the Interior Plains.

West of the Mississippi River lie the Great Plains. These are vast areas of grasslands. Further west, the land begins to rise, eventually leading to the **Rocky Mountains**. These mountains reach higher than 14,000 feet (4,267m). Along the crest of the Rocky Mountains is a ridge that divides North America's rivers. This is called a **continental divide**. Rivers east of the divide flow eastward, and rivers west of the divide flow westward.

> About how tall is the highest mountain in the Rockies?
> _______________________
> _______________________

Farthest west is the Pacific coast. The mountains here include the Cascade Range and the Sierra Nevada. Mount McKinley in Alaska is the highest mountain in North America.

> What is the highest mountain in North America?
> _______________________
> _______________________

## CLIMATE

The eastern United States is divided into three climate regions. The Northeast has a humid continental climate. To the south, the climate is humid subtropical. Farthest south, in the tip of southern Florida, the climate is tropical savanna.

> Circle the words that describe some of the different climates in the United States.

The climate in the Interior Plains varies. It is hot and dry in the Great Plains. But in most of the Midwest, the climate is humid continental. In the West, climates are mostly dry. Alaska has subarctic and tundra climates, while Hawaii is tropical.

## NATURAL RESOURCES

Our lives are affected in some way by natural resources every day. Much of our paper, food, gas, and electricity come from natural resources in the United States.

> Name three products that come from natural resources found in the United States.
> _______________________
> _______________________

## CHALLENGE ACTIVITY

**Critical Thinking: Drawing Inferences** Write three paragraphs describing what makes the physical geography of the United States so diverse.



 Interactive Reader and Study Guide

## The United States

**Section 2**

---

**MAIN IDEAS**

1. The United States is the world's first modern democracy.
2. The people and culture of the United States are very diverse.

# Key Terms and Places

**colony** a territory inhabited and controlled by people from a foreign land

**Boston** a major seaport in the British colonies during the mid-1700s

**New York** a major seaport in the British colonies during the mid-1700s

**plantation** a large farm that grows mainly one crop

**pioneers** the first settlers in the West

**bilingual** having the ability to speak two languages

# Section Summary

### FIRST MODERN DEMOCRACY

Europeans began settling in North America in the 1500s and setting up **colonies**. New cities such as **Boston** and **New York** became major seaports in the British colonies. Thousands of enslaved Africans were brought to the colonies and forced to work on **plantations**.

> **Name two major seaports in the British colonies.**
> ________________________
> ________________________

In 1774, many British colonists were unhappy with British rule. As a result, the Continental Congress adopted the Declaration of Independence in July 1776. To win independence, colonists fought the British in the Revolutionary War. The British were defeated in 1781 at the Battle of Yorktown in Virginia.

> **What happened in July 1776?**
> ________________________
> ________________________
> ________________________

After the war, the United States began to expand west. The first settlers were called **pioneers**.

The United States faced two world wars during the 1900s. After World War II, the United States and the Soviet Union became rivals in the Cold War, which lasted until the 1990s.



  Interactive Reader and Study Guide

The United States has a limited, democratic government based on the U.S. Constitution. The President and Congress are elected by U.S. citizens. Citizens may vote starting at age 18.

> **What kind of government does the United States have?**
>
> _______________________________
>
> _______________________________

## PEOPLE AND CULTURE

The majority of Americans are descendants of European immigrants. The United States is also home to people of many different cultures and ethnic groups. The United States is a diverse nation, where many languages are spoken, different religions are practiced, and a variety of foods are eaten.

For thousands of years, Native Americans were the only people living in the Americas. Descendents of enslaved Africans live throughout the country, with the highest population of African Americans living in the South. Other people migrated to the United States from different regions of the world. These include Asian countries, such as China, India, and the Philippines. Many Hispanic Americans originally migrated to the United States from Mexico, Cuba, and other Latin American countries.

> **Who were the first people who lived in the Americas?**
>
> _______________________________
>
> _______________________________

When people migrate to the United States, they bring parts of their culture with them, including their religions, food, and music. Some Americans are **bilingual**. Other than English, Spanish is the most widely spoken language in the United States.

People of different ethnic groups make the United States a very diverse country.

> **Underline the sentence that explains from where many Hispanic Americans originally migrated.**

## CHALLENGE ACTIVITY

**Critical Thinking: Drawing Inferences** Imagine that you are about to turn 18. Make a list of the responsibilities you have as a United States citizen.

# The United States

## Section 3

**MAIN IDEAS**

1. The United States has four regions—the Northeast, South, Midwest, and West.
2. The United States has a strong economy and a powerful military but is facing the challenge of world terrorism.

## Key Terms and Places

**megalopolis**  a string of large cities that have grown together

**Washington, D.C.**  the United States capital

**Detroit**  located in Michigan, and the nation's leading automobile producer

**Chicago**  the third-largest city in the United States and one of the busiest shipping ports on the Great Lakes

**Seattle**  Washington's largest city and home of a major software company

**terrorism**  violent attacks that cause fear

## Section Summary

### REGIONS OF THE UNITED STATES

Geographers often divide the United States into four main regions. These are the Northeast, the South, the Midwest, and the West.

The Northeast is the smallest region in the United States, as well as the most densely populated. Natural resources in the Northeast include rich farmland, coal, and fishing. Major seaports make it possible to ship products to markets around the world.

The Northeast is covered by a string of large cities called a **megalopolis**. It stretches from Boston to **Washington**, **D.C.** Other major cities are New York, Philadelphia, and Baltimore.

The South includes coastlines along the Atlantic Ocean and the Gulf of Mexico. The coastal plains provide farmers with rich soils for growing cotton, tobacco, and citrus fruits.

---

**What are the four main regions of the United States?**

_______________________

_______________________

_______________________

_______________________

**Name 5 cities that make up the megalopolis in the Northeast:**

_______________________

_______________________

_______________________

_______________________

_______________________

**Section 3, *continued***

Technology, education, and oil are also important industries in the South. Warm weather and beautiful beaches make tourism an important part of the South's economy.

> Underline the sentences that describe the industries that contribute to the South's economy.

The Midwest is one of the most productive farming regions in the world. Rich soils from the Mississippi River are perfect for raising livestock and producing corn, dairy products, and soybeans.

Most of the major cities in the Midwest, such as **Chicago** and **Detroit**, are located on rivers or the Great Lakes. This makes it easier to transport farm products, coal, and iron ore.

> Where are the major cities of the Midwest located?

The West is the largest region in the United States. California's mild climate and wealth of resources make it home to more than 10 percent of the country's population.

> What is the largest region in the United States?

Ranching, farming, coal, oil, gold, silver, copper, forestry, and fishing are important industries in the West. **Seattle**, Washington's largest city, is home to many of these industries.

## ECONOMY, MILITARY, AND TERRORISM

As the world's only superpower, the United States faces many issues today. Some of these include trade and **terrorism**. Trade with other countries is important to our nation's economy. Terrorism threatens our nation's safety. After the deadliest terrorist attack in U.S. history on September 11, 2001, the United States and other world leaders began working together to stop terrorism.

> What are two major issues the United States faces?

## CHALLENGE ACTIVITY

**Critical Thinking: Drawing Inferences** Now that you know about the physical features, climates, natural resources, industries, and economies of the regions in the United States, write about which region you would choose to live in and why.

# Canada

## CHAPTER SUMMARY

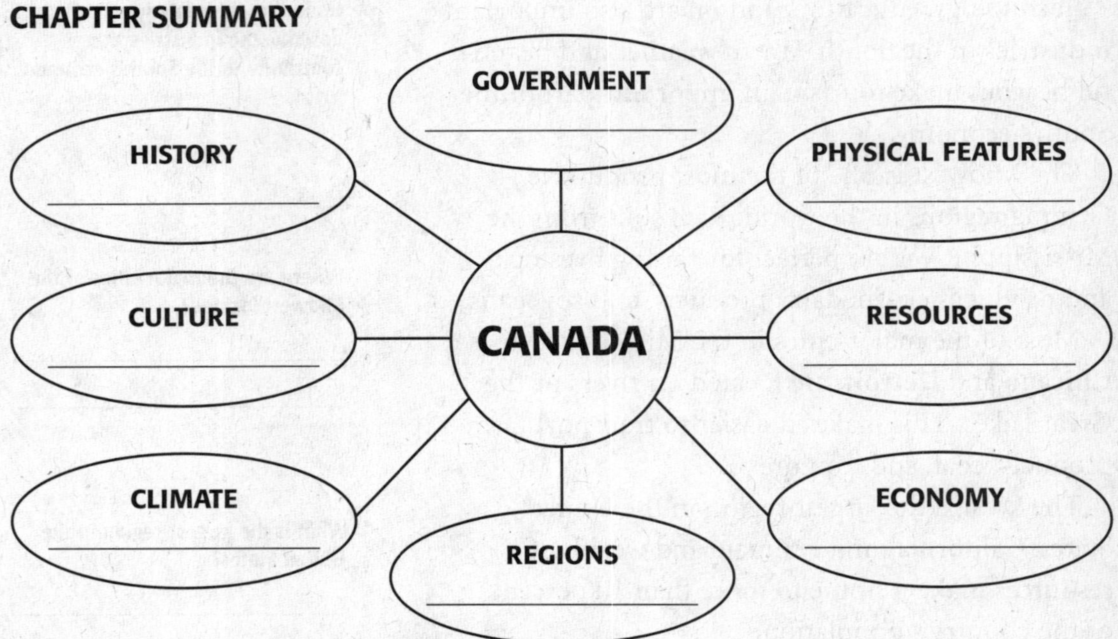


## COMPREHENSION AND CRITICAL THINKING

Use information from the graphic organizer to answer the following questions.

**1. Identify** Which aspects of Canadian life in the graphic organizer above are connected with physical features of the environment? Which aspects depend on human activities?

_______________________________________________________________

_______________________________________________________________

**2. Interpret** In which circle of the graphic organizer does each of these words best fit: trade, language, St. Lawrence River, minerals, parliament? Write each word in the correct circle.

**3. Draw Conclusions** Which aspects of Canadian life shown in the graphic organizer above have been directly affected by immigrants coming to Canada? Choose one and tell how it has been affected.

_______________________________________________________________

_______________________________________________________________

## Canada

**Section 1**

---

**MAIN IDEAS**

1. A huge country, Canada has a wide variety of physical features, including rugged mountains, plains, and swamps.
2. Because of its northerly location, Canada is dominated by cold climates.
3. Canada is rich in resources like fish, minerals, fertile soils, and forests.

# Key Terms and Places

**Rocky Mountains** mountains that extend north from the United States into western Canada

**St. Lawrence River** an important international waterway that links the Great Lakes to the Atlantic Ocean

**Niagara Falls** falls created by the waters of the Niagara River

**Canadian Shield** region of rocky uplands, lakes, and swamps

**Grand Banks** large fishing ground off the Atlantic coast near Newfoundland and Labrador

**pulp** softened wood fibers

**newsprint** cheap paper used mainly for newspapers

# Section Summary

### PHYSICAL FEATURES

Canada is the second-largest country in the world. Only Russia is larger. The United States and Canada share several physical features. Among these features are the Coast and **Rocky Mountains** as well as broad plains that stretch across both countries. The two nations also share a natural border formed by the **St. Lawrence River**.

**Niagara Falls** is another physical feature that the two countries share. The falls are created by the Niagara River as it drops over a rocky ledge.

One of Canada's unique features is the **Canadian Shield**. This rocky region covers about 1.8 million square miles of Canada. Canadian territory extends north to the Arctic Ocean where the land is covered with ice all year. Very few people live in this harsh environment.

> **How large is Canada?**
> _______________________
> _______________________

> **List three physical features that are in both Canada and the United States.**
> _______________________
> _______________________

**Section 1, *continued***

## CLIMATE

Canada's climate is greatly affected by its location. The country is far to the north of the equator. It is also at higher latitudes than the United States. Because of this, it has cold temperatures year-round.

The coldest part of Canada is close to the Arctic Circle. Central and northern Canada have subarctic climates. The far north has tundra and ice cap climates. About half of Canada has very cold climates.

> **How has Canada's location influenced its climate?**
> _______________________
> _______________________
> _______________________
> _______________________

## RESOURCES

Canada has many natural resources. These resources include fish, minerals, and forests. One of Canada's richest fishing areas is the **Grand Banks** near Newfoundland and Labrador, off the Atlantic coast. Large schools of fish once swam here, but too much fishing has reduced the number of fish in this part of the ocean.

> **Circle the names of three types of natural resources that Canada has.**
> _______________________
> _______________________

Canada has many mineral resources. It also has oil and gas. Many of Canada's mineral resources come from the Canadian Shield. Canada has more nickel, zinc, and uranium than any other country. Alberta produces most of Canada's oil and natural gas.

All across Canada, forests provide lumber and **pulp** to make paper. The United States, the United Kingdom, and Japan get much of their **newsprint** from Canada.

## CHALLENGE ACTIVITY

**Critical Thinking: Drawing Inferences** How are Canada's natural resources connected to its climate and physical features? Write two paragraphs explaining the connections.

# Canada

**Section 2**

**MAIN IDEAS**

1. Beginning in the 1600s, Europeans settled the region that would later become Canada.
2. Immigration and migration to cities have shaped Canadian culture.

## Key Terms and Places

**Quebec** province of Canada first settled by the French

**provinces** administrative divisions of a country

**British Columbia** province of Canada on the Pacific Coast

**Toronto** Canada's largest city

## Section Summary

### HISTORY

Indians and Inuit were the first people to live in Canada. The Cree Indians lived on the plains. They survived by hunting bison. The Inuit lived in the far north. They learned to survive in the harsh, cold climate by hunting seals, whales, and other animals.

The Vikings were the first Europeans to come to Canada, but they did not stay. European explorers and fishermen came in the late 1400s. Europeans soon began to trade with Native Canadians. The French built the first permanent settlements in what became Canada. They called the lands they claimed New France. In the 1700s Britain defeated France in the French and Indian War. Although the British took control, most French settlers stayed. Their way of life did not change much. Many descendants of these French settlers live in **Quebec** today.

The British divided their territory into two colonies called Upper and Lower Canada. The colony stayed divided until 1867 when the British passed a law making Canada a dominion, a territory under British influence. This act gave Canada its own government. The British also divided the country into **provinces**. In 1885,

> Circle the names of Canada's first people and Canada's first permanent European settlers.

> How did the French and Indian War affect the French colonists in New France?
>
> _______________________
>
> _______________________

> What is a dominion?
>
> _______________________
>
> _______________________

Canadians completed a railroad across Canada. It connected **British Columbia** with the eastern provinces. Canada increased in size by buying lands in the north where many Native Canadians live.

## CULTURE

Canadians come from many places. Some are descendants of early French and British settlers. In the late 1800s and early 1900s many immigrants came from Europe and the United States. Some hoped to find gold in Canada's Yukon Territory. Many Asian immigrants have come to Canada, especially from China, Japan, and India. Many Chinese immigrants came to work on the railroads. Immigrants helped Canada's economy grow. British Columbia was the first Canadian province to have a large Asian minority.

After World War II a new wave of immigrants came to large cities like **Toronto**. Many came from Asia. Others came from Europe, Africa, the Caribbean, and Latin America. In recent years, Canadians have also moved from farms in rural areas to large cities like Vancouver and Ontario.



## CHALLENGE ACTIVITY

**Critical Thinking: Making Inferences** How has immigration helped Canada's economy grow? Imagine that you are a person preparing to move to Canada. Write a letter to a friend explaining your reasons for moving and how you expect to contribute to the Canadian economy.

# Canada

**MAIN IDEAS**

1. Canada has a democratic government with a prime minister and a parliament.
2. Canada has four distinct geographic and cultural regions.
3. Canada's economy is largely based on trade with the United States.

## Key Terms and Places

**regionalism**  a strong connection that people feel toward their region

**maritime**  on or near the sea

**Montreal**  Canada's second largest city, one of the world's largest French-speaking cities

**Ottawa**  Canada's national capital

**Vancouver**  city on the Pacific coast with strong trade ties to Asia

## Section Summary

### CANADA'S GOVERNMENT

Canada's national government is like the U.S. government. Canada has a democratic central government led by a prime minister. This job is like that of a president. The prime minister also leads Parliament, Canada's governing body. Parliament is made up of the House of Commons and the Senate. Canadians elect members of the House of Commons. The prime minister appoints senators. Provincial governments are like our state governments. A premier leads each province.

> **Circle the title of the person who is the leader of Canada's government.**

### CANADA'S REGIONS

Canada has four regions. Each has its own cultural and physical features. In Quebec province, which is located in the Heartland region, **regionalism** has created problems between French and English speakers.

The eastern provinces are on the Atlantic coast. They include the **Maritime** Provinces—New Brunswick, Nova Scotia and Prince Edward Island—as well as Newfoundland and Labrador.



                     Interactive Reader and Study Guide

**Section 3, *continued***

Most people live in cities near the coast. Forestry and fishing are the major economic activities.

More than half of all Canadians live in the Heartland provinces of Ontario and Quebec. This region includes **Montreal**, Toronto, and **Ottawa**. Quebec is a center of French culture. Ontario is Canada's top manufacturing province. The Western Provinces include British Columbia on the Pacific coast and the prairie provinces—Manitoba, Saskatchewan, and Alberta. Farming is important there, especially growing wheat. In British Columbia, **Vancouver** is a major center for trade with Asia.

The Canadian North consists of the Yukon and Northwest territories and Nunavut, the Inuit homeland. The region is very cold and not many people live there. Nunavut has its own local government.

> Underline the Maritime and Heartland provinces. List two ways they are different from each other.
> ___________________________
> ___________________________

## CANADA'S ECONOMY

Many of Canada's economic activities are connected to its natural resources. Mining and manufacturing are key industries along with producing minerals. Most Canadians hold service jobs. Tourism is the fastest-growing service industry. Trade is also important. The United States is Canada's leading trading partner. The United States buys much of its lumber from Canada.

> Which type of economic activity employs the most workers?
> ___________________________
> ___________________________

## CHALLENGE ACTIVITY

**Critical Thinking: Drawing Conclusions** Why do you think Canada and the United States are such strong trading partners? What are the advantages and disadvantages of this strong trading relationship? Explain your answer in a one-page essay.

# Early History of Europe

## CHAPTER SUMMARY

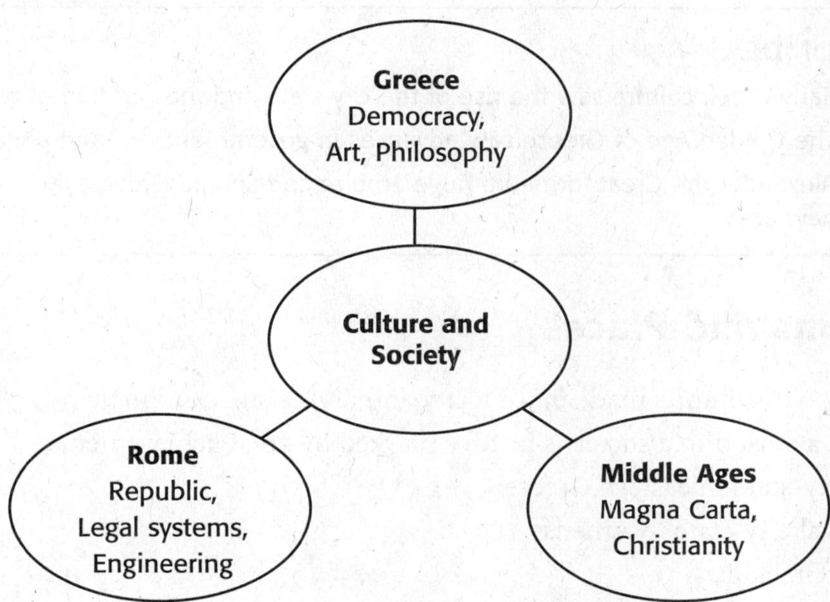


## COMPREHENSION AND CRITICAL THINKING

Use information from the graphic organizer to answer the following questions.

1. **Recall** Which culture(s) made major contributions to government? What were these contributions?

_______________________________________________________________

_______________________________________________________________

2. **Interpret** In which oval do the following words (*Gothic architecture, feudal system*) belong?

_______________________________________________________________

_______________________________________________________________

3. **Elaborate** Which cultural contributions could be added to both the "Greece" and "Rome" ovals?

_______________________________________________________________

_______________________________________________________________

# Early History of Europe

Section 1

**MAIN IDEAS**

1. Early Greek culture saw the rise of the city-state and the creation of colonies.
2. The Golden Age of Greece saw advances in government, art, and philosophy.
3. Alexander the Great formed a huge empire and spread Greek culture into new areas.

## Key Terms and Places

**city-states**  political units made up of a city and all the surrounding lands

**golden age**  a period in a society's history marked by great achievements

**Athens**  a city-state in eastern Greece

**Sparta**  a rival city-state to Athens

**Hellenistic**  Greeklike

## Section Summary

### EARLY GREEK CULTURE

To protect against invaders, early Greeks joined together. Over time, they developed into **city-states**, political units made up of a city and its surrounding lands. As populations grew, city-states formed colonies, or new cities—and Greek culture spread.

> Underline the reason why the Greeks established colonies.

### THE GOLDEN AGE OF GREECE

Greece is famous for its many contributions to world culture, especially during a period of great achievements called the **golden age**. Greece's golden age took place between 400 and 300 BC, after **Athens** and other city-states defeated a powerful Persian army around 500 BC. The defeat of the Persians increased the confidence of the Greeks, and they began to make many advances.

> Briefly define "golden age."
> ___________________________
> ___________________________

> What effect did the defeat of Persia have on the Greeks?
> ___________________________
> ___________________________

Athens became the cultural center of Greece during the golden age. Leaders such as Pericles supported the arts and other great works. But these leaders did not rule Athens, which became the world's first democracy. Instead, power was in



  Interactive Reader and Study Guide

**Section 1,** *continued*

the hands of the people, who voted in an assembly, which made the laws.

Many accomplishments in art, architecture, literature, philosophy, and science took place during Greece's golden age. For example, they were the first to create and perform drama, or plays. Greek philosophers such as Socrates, Plato, and Aristotle continue to shape how we think today.

> Underline the areas of accomplishment during the golden age.

Greece's golden age would not last. It came to an end when Athens and its rival **Sparta** went to war. Sparta had a strong army and was jealous of Athens's influence in Greece. The war raged for years, with other city-states joining in. Sparta finally won, but Greece overall had been weakened. It lay open to a foreign conqueror to take over.

## THE EMPIRE OF ALEXANDER

That conqueror was Alexander the Great, who took over Greece in the 330s BC. He conquered not only Greece, but also huge areas of the rest of the world from Greece to India, and most of central Asia. He dreamed of conquering more territory, but his tired and homesick troops refused to continue. Alexander died on his return home, at the age of 33.

> Do you think Alexander could have taken over Greece if Sparta and Athens had not gone to war? Why or why not?
>
> _______________________
> _______________________
> _______________________

Alexander admired Greek culture and wanted it to spread throughout his empire. He urged Greek people to move to new cities. Many Greeks did move. Greek culture then blended with other cultures. These blended cultures are referred to as **Hellenistic**, or Greeklike.

> What was Alexander's attitude toward Greek culture?
>
> _______________________
> _______________________

## CHALLENGE ACTIVITY

**Critical Thinking: Analyzing Information** Imagine that you are a newspaper reporter from a Greek city-state, and your assignment is to write about Athens. Write a brief article describing what you have seen on your visit. In your article include information about at least one of Athens' famous artists, architects, scientists, philosophers, or writers.



     Interactive Reader and Study Guide

# Early History of Europe

**Section 2**

### MAIN IDEAS

1. The Roman Republic was governed by elected leaders.
2. The Roman Empire was a time of great achievements.
3. The spread of Christianity began during the empire.
4. Various factors helped bring about the decline of Rome.

## Key Terms and Places

**Rome**  a city in Italy

**republic**  a type of government in which people elect leaders to make laws for them

**Senate**  a council of rich and powerful Romans who helped run the city

**citizens**  people who could take part in the Roman government

**Carthage**  a city in North Africa

**empire**  a land that includes many different peoples and lands under one rule

**aqueducts**  channels used to carry water over long distances

## Section Summary

### THE ROMAN REPUBLIC

**Rome** began as a small city in Italy. It was ruled by a series of kings, some of whom were cruel rulers. Over time the Romans formed a **republic**, in which elected leaders made all government decisions. The leaders worked with the **Senate**, a group of powerful men. **Citizens** voted and ran for office.

> What is a republic?
> ______________________
> ______________________

Before long, Rome began to expand its territory. It took over much of the Mediterranean world, including the city of **Carthage** in North Africa. A general named Julius Caesar conquered many new lands. Afraid of his power, a group of Senators killed him in 44 BC.

### THE ROMAN EMPIRE

Caesar's adopted nephew Octavian became Rome's first emperor, ruling a huge **empire**. This far-flung land contained many different people. Octavian was also called Augustus, or "honored one," because of

> Circle the name of Rome's first emperor.



  Interactive Reader and Study Guide

**Section 2, *continued***

his many accomplishments. Augustus conquered many new lands. He built monuments and roads.

The Romans had a long period of peace and achievement called the Pax Romana. Their many building projects included **aqueducts**—channels to carry water long distances. They wrote literature and built legal systems that have had a worldwide influence. The founders of the United States used the Roman government as a model for our own.

> Name a country whose legal system was influenced by the Romans.
>
> _______________________

## THE SPREAD OF CHRISTIANITY

Christianity first appeared in the Roman Empire. The religion was based on the teachings of Jesus of Nazareth. Jesus's followers preached throughout the Roman world. For many years, Rome tried unsuccessfully to stop the spread of Christianity. Then in the 300s an emperor named Constantine became a Christian. Christianity soon became Rome's official religion. By the end of the 300s it had become a powerful force in the Roman world.

> Underline the name of the Roman emperor who converted to Christianity.

## THE DECLINE OF ROME

Rome's decline had several causes. A number of bad emperors ignored their duties and the needs of the Roman people. Military leaders tried to take over, but many of them were poor rulers as well.

The emperor Constantine created a new capital in a central location. But this change was not enough. The empire, weakened by its internal problems, was vulnerable to invaders called barbarians. In 476 they attacked Rome and removed its emperor. The Roman Empire was no more.

> Why is the year 476 considered the end of the Roman Empire?
>
> _______________________

## CHALLENGE ACTIVITY

**Critical Thinking: Analyzing Information** Imagine that you are a Roman citizen. Write a letter to either Constantine or Augustus, asking him to take an important action to improve the Roman Empire. Include reasons why this action is needed, based on details in the section.

# Early History of Europe

Section 3

**MAIN IDEAS**

1. The Christian church influenced nearly every aspect of society in the Middle Ages.
2. Complicated political and economic systems governed life in the Middle Ages.
3. The period after 1000 was a time of great changes in medieval society.

## Key Terms and Places

**Middle Ages**  a period of history between ancient and modern times that lasted from about 500 until about 1500

**pope**  the head of the Christian church

**Crusade**  a religious war

**Holy Land**  the region in which Jesus had lived

**Gothic architecture**  a style known for its high pointed ceilings, tall towers, and stained glass windows

**feudal system**  a system of exchanging land for military service

**manor**  a large estate owned by a noble or a knight

**nation-state**  a country united under a single strong government

## Section Summary

### THE CHRISTIAN CHURCH AND SOCIETY

After the late 400s, Europe broke into many small kingdoms. The period from 500 until about 1500 is called the **Middle Ages**. During this time, no one leader could unify Europe as the Romans had done. As a result, the Christian church gained influence, and church leaders became powerful.

> Circle the dates when the Middle Ages took place.

The **pope** was the head of the church. One pope started a religious war called a **Crusade**. He wanted Europeans to take over the **Holy Land**, where Jesus had lived. It was then in the hands of the Muslims. The Crusaders failed. However, they brought back new foods, goods, and ideas. Trade between Europe and Asia increased.

> Who was the leader of the Christian church?
>
> _______________________

The church had a major influence on art and architecture. Many churches built during this period are examples of **Gothic architecture**. Most people's lives centered around their local church.



  Interactive Reader and Study Guide

## LIFE IN THE MIDDLE AGES

Religion was not the only influence on people's lives. Two other major influences were the **feudal system** and the **manor** system.

The feudal system was mainly a relationship between nobles and knights. The nobles gave land to knights. In turn, the knights promised to help defend their lands and the king.

The manor system was a relationship between manor, or large estate, owners and workers. The owners provided workers with a place to live and a piece of land where they could grow their own food. In exchange, most of the crops went to the owners.

> **Who participated in the feudal system? In the manor system?**
>
> _________________________
>
> _________________________

## CHANGES IN MEDIEVAL SOCIETY

France's William the Conqueror invaded England in 1066. He became king, and built England's first strong government. In 1215, however, the English king lost some of his power. A group of nobles insisted that the king should not be above the law. The nobles drew up a document called Magna Carta, which limited the King's power.

> **How did Magna Carta affect the king?**
>
> _________________________

In 1347 a disease called the Black Death swept through Europe. It killed about a third of the population. The plague caused a labor shortage. As a result, people could demand higher wages.

> **What is one negative and one positive outcome of the Black Death?**
>
> _________________________
>
> _________________________

In 1348 the Hundred Years' War broke out between England and France. The French won, and kings began working to end the feudal system and gain more power. France became a **nation-state**, a country united under a single government. Other nation-states arose around Europe, and the Middle Ages came to an end.

## CHALLENGE ACTIVITY

**Critical Thinking: Making Generalizations** Draw up a document like Magna Carta that includes your ideas of the basic rights for common people as well as limits on the power of rulers.

# History of Early Modern Europe

## CHAPTER SUMMARY

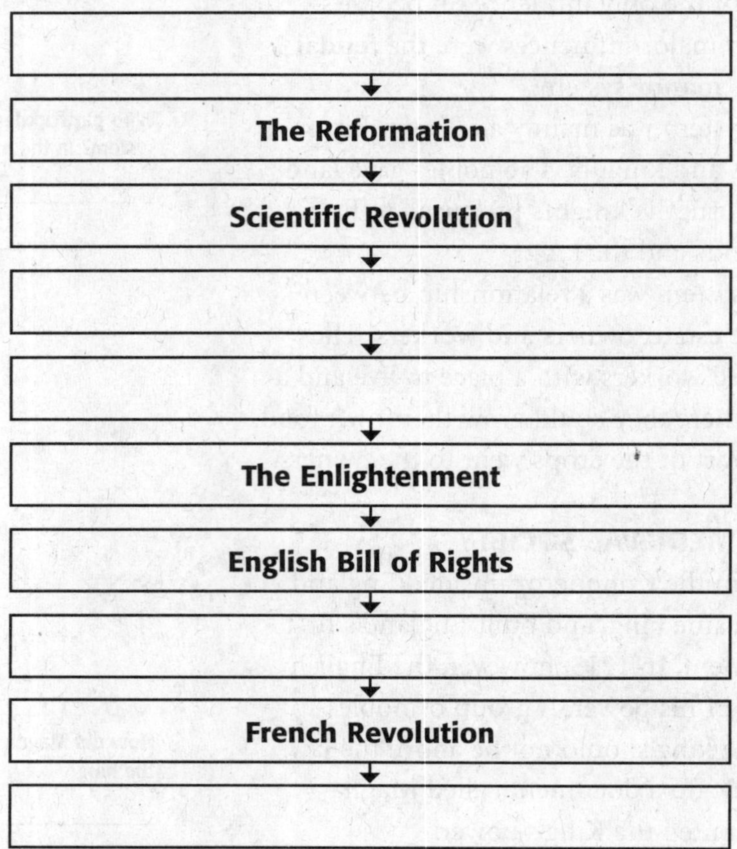


## COMPREHENSION AND CRITICAL THINKING

Use the graphic organizer to answer the following questions.

**1. Sequence** Place these events in the correct order on the chart: *The American Revolution, Voyages of Discovery, The Renaissance, The Industrial Revolution, New Empires*

**2. Explain** Pick an event on the chart and explain how it is connected to the Enlightenment.

_______________________________________________________________

_______________________________________________________________

**3. Compare/Contrast** How were the Scientific Revolution and the Enlightenment similar? How did they differ?

_______________________________________________________________

_______________________________________________________________

# History of Early Modern Europe

## Section 1

**MAIN IDEAS**
1. The Renaissance was a period of new learning, new ideas, and new advances in art, literature, and science.
2. The Reformation changed the religious map of Europe.

# Key Terms and Places

**Renaissance** period of creativity and new ideas that swept Europe from about 1350 through the 1500s

**Florence** Italian city that became rich through trade during and after the Crusades

**Venice** Italian city that became rich through trade during and after the Crusades

**humanism** new way of thinking and learning that emphasized the abilities and accomplishments of human beings

**Reformation** religious reform movement that began with complaints about problems within the Catholic Church

**Protestants** Christians who split from the Catholic Church over religious issues

**Catholic Reformation** series of reforms launched by Catholic Church officials

# Section Summary

## THE RENAISSANCE

The **Renaissance** started in Italy in such cities as **Florence** and **Venice**. These cities became rich through trade. As goods from Asia moved through these cities, Italians became curious about the larger world. At this same time, scholars from other parts of the world came to Italy, bringing books written by ancient Greeks and Romans. Interest in Greece and Rome grew. People studied subjects once taught in Greek and Roman schools such as history, poetry, and grammar. These subjects are known as the humanities. Increased study of the humanities led to **humanism**—an idea that people are capable of great achievements.

Renaissance artists developed new painting techniques such as perspective, which made their art look more realistic. The artists Michelangelo and Leonardo da Vinci showed their belief in humanism by making the people in their paintings

> Circle the name of the country where the Renaissance began.

> List three developments that led to the Renaissance.
>
> _______________________
>
> _______________________
>
> _______________________
>
> _______________________
>
> _______________________

> Circle the names of three Renaissance artists or writers.



 Interactive Reader and Study Guide

look like unique individuals. William Shakespeare's plays looked closely at human nature and behavior.

Reading about Greek and Roman scientific advances inspired Europeans to study math, astronomy, and other sciences. Some used their new knowledge to create new inventions. Johannes Gutenberg's invention, the movable-type printing press, printed books quickly and cheaply. It spread Renaissance ideas to all parts of Europe.

> **What effect did Gutenberg's printing press have on the Renaissance?**
> _______________________
> _______________________

## THE REFORMATION

The **Reformation**, a religious reform movement, began in Germany. Many people there felt priests and other Catholic Church officials cared more about power than their religious duties. Martin Luther, a German monk, was one of the first people to protest against the church. In 1517 he nailed a list of complaints on a church door in Wittenberg. Angry church officials expelled him from the church. Luther's followers became the first **Protestants**, splitting off from the Catholic Church. Other reformers created their own churches. By 1600 many Europeans had become Protestants.

> **Underline the sentence that tells how Luther made his protest.**

In response, Catholic Church leaders began a series of reforms, called the **Catholic Reformation**. They asked churches to focus more on religious matters. They tried to make church teachings easier to understand. Priests and teachers went to Asia, Africa, and other lands to spread Catholic teachings. After the Reformation, religious wars broke out in Europe between Catholics and Protestants. These religious wars led to many changes in Europe.

## CHALLENGE ACTIVITY

**Critical Thinking: Analyzing** What events prompted Catholic leaders to begin the Catholic Reformation?

# History of Early Modern Europe

**Section 2**

**MAIN IDEAS**

1. During the Scientific Revolution, discoveries and inventions expanded knowledge and changed life in Europe.

2. In the 1400s and 1500s, Europeans led voyages of discovery and exploration.

3. As Europeans discovered new lands, they created colonies and new empires all over the world.

## Key Terms

**Scientific Revolution** series of events that led to the birth of modern science
**New World** the land discovered by Columbus that came to be known as America
**circumnavigate** travel all the way around Earth

## Section Summary

### THE SCIENTIFIC REVOLUTION

Before the 1500s, most educated people depended on authorities such as ancient Greek writers and church officials for information. During the **Scientific Revolution**, people began to believe that what they observed was more important than what they were told. They developed logical explanations for how the world worked, based on what they observed. This focus on observation marked the start of modern science.

Many Europeans feared the spread of scientific ideas. Church officials opposed many ideas because they went against church teachings. For example, the church taught that the sun circled the Earth. From studying the skies, scientists now thought that the Earth circled the sun. In 1632 church officials arrested Italian scientist Galileo for publishing a book that supported this view. However, science still developed rapidly.

New discoveries occurred in astronomy, biology, physics, and other fields. New inventions included the telescope, microscope, and thermometer.

> Circle two ways Europeans learned about the world before the Scientific Revolution.

> Why did Church officials fear the spread of scientific ideas?
> _______________________
> _______________________

Isaac Newton made one of the most important discoveries—explaining how gravity works.

## THE VOYAGES OF DISCOVERY

With improved devices such as the compass and astrolabe, and better ships, Europeans made longer, safer sea voyages. They found new routes to distant places. Europeans had many reasons for exploring. Some were curious about the world; some wanted riches, fame, adventure; and others hoped to spread Christianity. Prince Henry of Portugal sought new trade routes to Asia. After many other voyages, Portuguese explorer Vasco da Gama succeeded in finding a water route to Asia.

The voyages of Christopher Columbus, paid for by Queen Isabella of Spain, were the most significant. His explorations of the **New World** led other European countries to explore this new land, too. The crew of Ferdinand Magellan were the first people to **circumnavigate**, or sail completely around, the world. Others followed. Voyages of discovery added greatly to European wealth—and people's knowledge of the world.

## NEW EMPIRES

The Spanish were the first Europeans to build colonies in the Americas. They used metal weapons unknown to native peoples to conquer and destroy the Aztec and Inca empires. Deadly diseases carried by the Spanish also helped them conquer new lands. The gold the Spanish found in the Americas made Spain the richest country in Europe. Britain, France, and other European nations also founded colonies in the Americas. In many places they forced out Native Americans. European nations grew rich from trade in wood, fur, and other natural resources.

## CHALLENGE ACTIVITY

**Critical Thinking: Drawing Conclusions** What were three results of European explorations?

> Circle the three advances that helped make the voyages of discovery possible.

> List three explorers and where they explored.
>
> ___________________
> ___________________
> ___________________

> What effect did European explorers and colonists have on Native Americans?
>
> ___________________
> ___________________

# History of Early Modern Europe

Section 3

---

**MAIN IDEAS**

1. During the Enlightenment, new ideas about government took hold in Europe.
2. The 1600s and 1700s were an Age of Revolution in Europe.
3. Napoleon Bonaparte conquered much of Europe after the French Revolution.

## Key Terms

**Enlightenment** period in the 1600s and 1700s when the use of reason shaped European ideas about society and politics, also known as the Age of Reason

**English Bill of Rights** 1689 document listing rights of Parliament and the English people

**Declaration of Independence** document signed in 1776 that declared the American colonies' independence from Britain

**Declaration of the Rights of Man and of the Citizen** French constitution that guaranteed some rights of French citizens and made taxes fairer

**Reign of Terror** period of great violence during the French Revolution

## Section Summary

### THE ENLIGHTENMENT

During the **Enlightenment** many people questioned common ideas about politics and government. However, most of Europe was ruled by kings and queens, also called monarchs. Most monarchs believed God gave them the right to rule as they chose. This belief was called rule by divine right.

Enlightenment thinkers disagreed. John Locke saw government as a contract, or binding legal agreement, between a ruler and the people. A ruler's job was to protect people's rights. If a ruler did not do this, people had the right to change rulers. Jean-Jacques Rousseau also felt government's purpose was to protect people's freedoms. Such ideas inspired revolutions and political change.

> Underline the sentence that explains what rule by divine right is.

> Circle the names of two Enlightenment thinkers who had new ideas about the rights of citizens and the role of government.

---



 Interactive Reader and Study Guide

**Section 3,** *continued*

## THE AGE OF REVOLUTION

In the 1600s, England's rulers fought with Parliament for power. As a result, in 1689 Parliament passed the **English Bill of Rights** and made the king agree to honor the Magna Carta. These steps limited the monarch's power and gave more rights to Parliament and the English people.

Enlightenment ideas spread to Britain's North American colonies. There, colonial leaders claimed Britain had denied their rights—and started the American Revolution. In July of 1776 Americans signed the **Declaration of Independence**, declaring freedom for the American colonies.

The American victory inspired the French people to fight for their rights. Members of the Third Estate, France's largest and poorest social class, demanded a part in government. They formed the National Assembly and demanded that the king limit his powers. When he refused, the French Revolution began. The National Assembly then issued the **Declaration of the Rights of Man and of the Citizen**, a new constitution for France.

France's revolutionary leaders ended the monarchy, but soon after the **Reign of Terror** began. It was a very violent time. After it ended, people longed for a strong leader to restore order.

> Why is this period called the Age of Revolution?

> Circle the names of documents that changed how rulers governed and gave citizens more rights.

## NAPOLEON BONAPARTE

In 1799 a smart military leader named Napoleon Bonaparte took control of France. He led his armies, conquering most of Europe and building a French empire. Napoleon created a fair legal system, called the Napoleonic Code, but he was a harsh ruler. Napoleon's armies were defeated in 1814 and 1815. Soon after, European leaders met to decide how to divide up the former French empire.

> List two of Napoleon's achievements.

## Challenge Activity

**Critical Thinking: Drawing Conclusions** In the Age of Revolution, were ideas or armies more important? Give support for your answer.



         Interactive Reader and Study Guide

**Section 4**

**MAIN IDEAS**

1. Britain's large labor force, raw materials, and money to invest led to the start of the Industrial Revolution.

2. Industrial growth began in Great Britain and then spread to other parts of Europe.

3. The Industrial Revolution led to both positive and negative changes in society.

## Key Terms

**Industrial Revolution** period of rapid growth in machine-made goods

**textiles** cloth products

**capitalism** an economic system in which individuals own most businesses and resources, and people invest money in hopes of making a profit

**suffragettes** women who campaigned for the right to vote

## Section Summary

### START OF THE INDUSTRIAL REVOLUTION

Changes in agriculture helped prepare Britain for industrial growth. Rich farmers bought land and created larger, more efficient farms. At the same time, Europe's population grew, creating a need for more food. To meet this need, farmers tried new farming methods and invented new machines.

These improved methods and inventions helped farmers grow more crops, but with fewer workers. As a result, many small farmers and farm workers lost their farms and jobs—and moved to the cities.

In Britain, all these changes sparked the **Industrial Revolution**. By the 1700s, Britain had labor, natural resources, and money to invest—all the resources needed for industry to grow. Demand for manufactured goods soon grew. People looked for ways to make these goods even faster.

> Underline the sentence that tells why farmers needed to grow more food.

> Circle three resources needed for industrial growth.

Section 4, *continued*

## INDUSTRIAL GROWTH

The **textile** industry, makers of cloth products, developed first. In the early 1700s, cloth was still made by hand. This changed in 1769 when Richard Arkwright invented a waterpowered spinning machine. Other new machines enabled workers to make large amounts of cloth. As a result, the price of cloth fell. Soon workers were using machines to make many other kinds of goods.

Most early machines relied on water power. Factories, the buildings that housed the machines, had to be built near rivers. In the 1760s James Watt built the first modern steam engine. Factories could now be set up in cities. In 1856 Henry Bessemer invented a new way to make steel, and the steel industry grew. Transportation became faster as steam engines powered boats and trains.

Industrial growth changed how people worked. Many people—including children and young women—worked in unsafe factories. They worked long hours, usually for poor wages. However, by the late 1800s, the Industrial Revolution had spread. Industrial growth resulted in a new economic system—**capitalism**, in which individuals own most businesses and resources.

## CHANGES IN SOCIETY

The Industrial Revolution made life better for some, but worse for others. Manufactured goods became cheaper. New inventions made life easier. More people joined the middle class. Meanwhile, cities grew, becoming dirty, noisy, and crowded. Workers often remained poor, living in unsafe apartments where diseases spread. Some women, called **suffragettes**, pressed for the right to vote. Other reformers worked to improve society, too.

## CHALLENGE ACTIVITY

**Critical Thinking: Drawing Conclusions** Did the benefits of the Industrial Revolution outweigh the problems it caused? Explain.

> Circle the industry that developed first. Then underline the invention that helped it grow first.

> Why do you think the price of cloth fell as workers made more cloth?
> _______________________
> _______________________

> Explain why you think the Industrial Revolution led to a new economic system.
> _______________________
> _______________________

> Underline how people's lives changed as a result of the Industrial Revolution.

# Modern European History

## CHAPTER SUMMARY

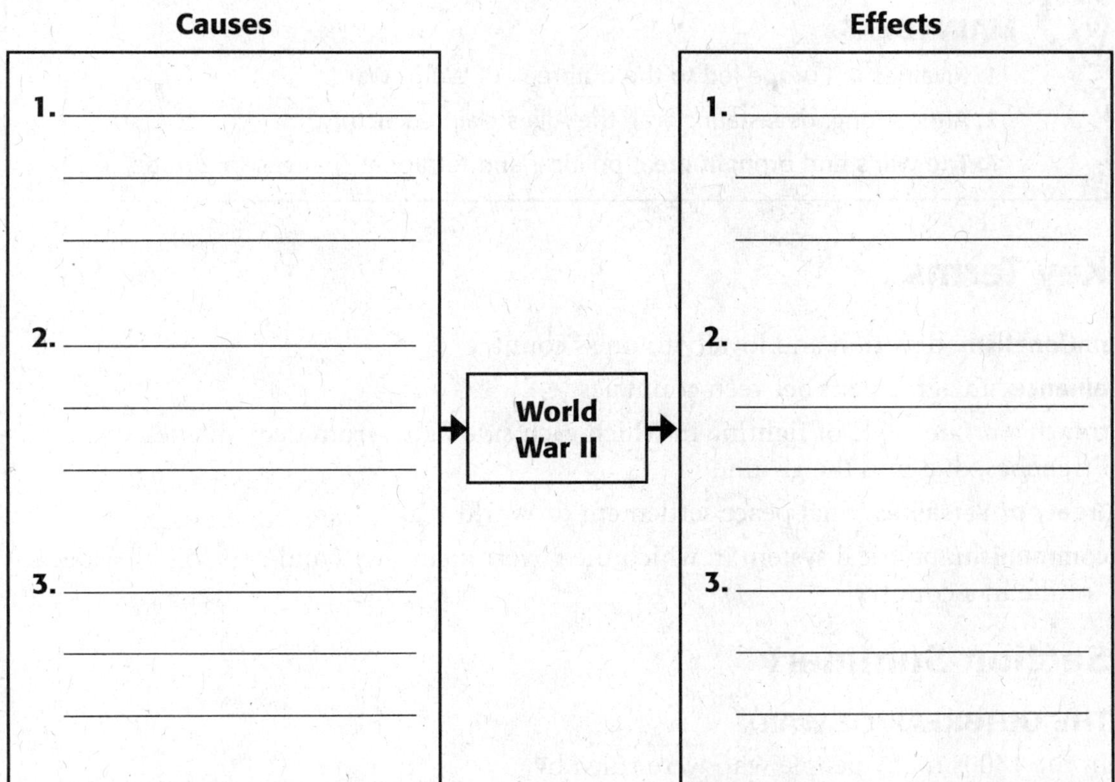


## COMPREHENSION AND CRITICAL THINKING

Use the graphic organizer to answer the following questions.

1. **Identify** What were the three main causes of World War II? Choose from the following: *Great Depression, United Nations, assassination of Archduke Ferdinand, atomic bomb, Germany's invasion of Poland, European dictatorships.* Write your answer in the spaces provided in the graphic organizer.

2. **Identify** What were the three main effects of World War II? Choose from the following: *Treaty of Versailles, Cold War, Poland gaining independence, communists taking control of Russia, the division of Europe, United Nations.* Write your answer in the spaces provided in the graphic organizer.

3. **Predict** Will Europeans continue to cooperate politically and economically?

_________________________________________________________________

_________________________________________________________________

_________________________________________________________________



          Interactive Reader and Study Guide

# Modern European History

### Section 1

**MAIN IDEAS**

1. Rivalries in Europe led to the outbreak of World War I.
2. After a long, devastating war, the Allies claimed victory.
3. The war's end brought great political and territorial changes to Europe.

## Key Terms

**nationalism**  devotion and loyalty to one's country

**alliance**  an agreement between countries

**trench warfare**  style of fighting in which each side fights from deep ditches, or trenches, dug into the ground

**Treaty of Versailles**  final peace settlement of World War I

**communism**  political system in which the government owns and controls all aspects of life in a country

## Section Summary

### THE OUTBREAK OF WAR

In the 1800s many people who were ruled by empires wanted to form their own nations. As **nationalism**—devotion and loyalty to one's country—became more common, tensions grew. In 1882 Italy, Germany, and Austria-Hungary formed the Triple Alliance. This **alliance** was an agreement to fight together if any of the three was attacked. Britain, Russia, and France also formed an alliance, the Triple Entente. By the 1900s many countries were preparing for war by building up their armies and stockpiling weapons. Germany and Great Britain built strong navies and powerful new battleships.

> Underline the sentence that explains what an alliance is.

One source of tension was Bosnia and Herzegovina, a province of Austria-Hungary that Serbia wanted to control. On June 28, 1914, Archduke Francis Ferdinand of Austria-Hungary was killed by a Serbian gunman. Serbia turned to Russia for help, and the alliance system split Europe into warring sides.

> When was Archduke Francis Ferdinand killed?
> ____________________



       Interactive Reader and Study Guide

**Section 1, *continued***

## WAR AND VICTORY

Germany attacked the Allies (France, Great Britain, Serbia, and Russia), sending its army into Belgium and France. Russia attacked the Central Powers (Germany and Austria-Hungary) from the east. The two sides quickly prepared for **trench warfare** by digging hundreds of miles of trenches, which were easy to defend but hard to attack. Millions of soldiers died in the trenches. New weapons, such as machine guns, poison gas, and tanks, were designed to break into the trenches.

> Circle the names of two countries that were part of the Central Powers.

In 1917 German U-boats attacked American ships that were helping Britain. The United States entered the war, strengthening the Allies. Around the same time, Russia pulled out of the war, and Germany attacked the Allies. This last effort failed, and the Central Powers surrendered in 1918.

> Why did the United States enter the war?
>
> ___________________________
> ___________________________

## THE WAR'S END

World War I, in which over 8.5 million soldiers died, changed Europe forever. American president Woodrow Wilson wanted a just peace after the war, but the **Treaty of Versailles** blamed Germany alone for the war. The Germans were forced to slash the size of their army and pay billions of dollars. Revolutions following the war replaced the German emperor with a fragile republic and the Russian czar (emperor) with a Communist leader, Vladimir Lenin. **Communism** is a political system in which the government owns and controls every aspect of life. In other government changes, Austria and Hungary became separate countries, Poland and Czechoslovakia gained independence, Yugoslavia was formed, and Finland, Latvia, Lithuania, and Estonia broke away from Russia.

> Where was the first communist government established?
>
> ___________________________

## CHALLENGE ACTIVITY

**Critical Thinking: Draw Inferences** Write a sentence explaining why machine guns and trench warfare were such a deadly combination.



                                      Interactive Reader and Study Guide

# Modern European History

### Section 2

## MAIN IDEAS

1. Economic and political problems troubled Europe in the years after World War I.
2. World War II broke out when Germany invaded Poland.
3. Nazi Germany targeted the Jews during the Holocaust.
4. Allied victories in Europe and Japan brought the end of World War II.

## Key Terms

**Great Depression**  a global economic crisis in the 1930s

**dictator**  a ruler who has total control

**Axis Powers**  an alliance among Germany, Italy, and Japan

**Allies**  France, Great Britain, and other countries that opposed the Axis

**Holocaust**  the attempt by the Nazi government during World War II to eliminate Europe's Jews

## Section Summary

### PROBLEMS TROUBLE EUROPE

With the economy booming after World War I, the United States provided loans to help Europe rebuild. But the U.S. stock market crashed in 1929, starting a global economic crisis, the **Great Depression**. Without United States funds, banks failed in Europe, and many people lost their jobs.

> **What caused many Europeans to lose their jobs after 1929?**
> _______________________________
> _______________________________

People blamed their leaders for hard times. In some countries, **dictators** gained power by making false promises. In the 1920s, Benito Mussolini spoke of bringing back the glory of the Roman Empire to Italy, but instead took away people's rights. In Russia in 1924, Joseph Stalin became dictator and oppressed the people, using secret police to spy on them. Promising to restore Germany's military and economic strength, Adolf Hitler rose to power in 1933. Once in power, he outlawed all political parties except the Nazi party. He also discriminated against Jews and other groups he believed inferior.

> **Underline the sentence that tells Adolf Hitler's actions when he first took power.**



 Interactive Reader and Study Guide

**Section 2, *continued***

## WAR BREAKS OUT

No one moved to stop Mussolini when he invaded Ethiopia in 1935. Meanwhile, Hitler added Austria and Czechoslovakia to the German empire in 1938. When Hitler invaded Poland in 1939, France and Britain declared war on Germany, beginning World War II. Germany, Italy, and Japan formed the **Axis Powers**. They were opposed by the **Allies**—France, Great Britain, and others. The Axis won most of the early battles and soon defeated France. But Britain withstood intense bombing and did not surrender. The German army then turned toward Eastern Europe and the Soviet Union, and Italy invaded North Africa. In 1941 Japan attacked the United States at Pearl Harbor, Hawaii.





## THE HOLOCAUST

The **Holocaust** was the Nazi party's plan to eliminate people they believed were inferior, especially the Jews. The Nazis started by taking away the rights of Jews, and many fled the country. By 1942 the Nazis had put millions of Jews in death camps, such as the one in Auschwitz, Poland. Some Jews tried to hide, escape, or fight back. In the end, however, the Nazis murdered some 6 million Jews and several million non-Jews.

## END OF THE WAR

In 1943 the Allies won some key battles. Then on "D-Day" in 1944 they invaded Normandy, France, and paved the way for an advance toward Germany. The war ended in 1945 soon after the U.S. dropped an atomic bomb on Japan. The war took the lives of more than 50 million people and led to the formation of the United Nations. It also made the United States and the Soviet Union the world's most powerful countries.



## CHALLENGE ACTIVITY

**Critical Thinking: Sequence** Make a time line showing the main events leading up to Germany's invasion of Poland in 1939.



 Interactive Reader and Study Guide

# Modern European History

Section 3

### MAIN IDEAS

1. The Cold War divided Europe between democratic and Communist nations.
2. Many Eastern European countries changed boundaries and forms of government at the end of the Cold War.
3. European cooperation has brought economic and political change to Europe.

## Key Terms

**superpowers**  strong and influential countries

**Cold War**  period of tense rivalry between the United States and the Soviet Union

**arms race**  competition between countries to build superior weapons

**common market**  group of nations that cooperates to make trade among members easier

**European Union (EU)**  organization that promotes political and economic cooperation in Europe

## Section Summary

### THE COLD WAR

After World War II, the two **superpowers**—the United States and the Soviet Union—distrusted each other. This led to the **Cold War**, a period of tense rivalry between these two countries. The Soviet Union stood for communism, and the United States stood for democracy and free enterprise.

> Underline the sentence that defines the Cold War.

The United States and several Western nations formed an alliance called the North Atlantic Treaty Organization (NATO). The Soviet Union and most Eastern European countries were allies under the Warsaw Pact. The two sides used the threat of nuclear war to defend themselves.

> Which alliance did most Eastern European countries join after World War II?
>
> _______________________

Germany split into East Germany and West Germany. Communist leaders built the Berlin Wall to prevent East Germans from fleeing to the West. Western countries were more successful economically than Communist Eastern Europe. People in the East suffered from shortages of money, food, clothing, and cars.





Interactive Reader and Study Guide

## THE END OF THE COLD WAR

By the 1980s, the **arms race**—a competition to build superior weapons—between the Soviet Union and the U.S. was damaging the Soviet economy. To solve the problem, Soviet leader Mikhail Gorbachev made changes. He reduced government control of the economy and held democratic elections.

These policies helped inspire change throughout the East. Poland and Czechoslovakia threw off Communist rule. The Berlin Wall came down in 1989. East and West Germany reunited to form a single country again in 1990. In December 1991 the Soviet Union broke up.



Ukraine, Lithuania, and Belarus became independent countries. In 1993 Czechoslovakia split peacefully into the Czech Republic and Slovakia. However, ethnic conflict in Yugoslavia caused much violence. By 1994 Yugoslavia had split into five countries—Bosnia and Herzogovina, Croatia, Macedonia, Slovenia, and Serbia and Montenegro.



## EUROPEAN COOPERATION

After two deadly wars, many Europeans thought a sense of community would make more wars less likely. In the 1950s West Germany, Luxembourg, Italy, Belgium, France, and the Netherlands moved toward unity with a **common market**, a single economic unit to improve trade between members. Today 25 countries have joined to make up the **European Union (EU)**, and many use a common currency, the euro. The European Union deals with issues such as environment, trade, and migration. Their governing body has executive, legislative, and judicial branches. Representatives are selected from all member nations. The EU has helped unify Europe, and other countries hope to join in the future.



**Critical Thinking: Cause and Effect** How might the establishment of the European Union make wars in Europe less likely in the future?

# Southern Europe

## CHAPTER SUMMARY

### Southern Europe

| Physical Features | Greece |
|---|---|
| • Iberian, Italian, and Balkan peninsulas<br>• Pyrenees, Alps, Apennines, and Pindus<br>• coastal plains and river valleys<br>• Mediterranean and other seas; rivers<br>• Mediterranean climate is a resource.<br>• The seas are a resource. | • ancient advances in many fields<br>• under foreign rule till 1800s<br>• Religion and family are central to life.<br>• Athens is both modern and ancient.<br>• Rural people still work on farms.<br>• Tourism and shipping are important. |
| **Italy** | **Spain and Portugal** |
| • Roman Empire was vast and powerful.<br>• Renaissance inspired great creativity.<br>• Nationalism made unification possible.<br>• Italy is a trendsetter in many fields.<br>• North is stronger economically.<br>• South receives incentives for industry. | • Foreign rulers included the Moors.<br>• Both established and lost colonies.<br>• Basques have separate culture.<br>• Spain: flamenco; Portugal: fados<br>• buildings influenced by Muslim design<br>• parliamentary monarchy; republic |

## COMPREHENSION AND CRITICAL THINKING

Use information from the graphic organizer to answer the following
questions.

**1. Explain** Why are southern Europe's seas and climate valuable resources?

_______________________________________________

**2. Identify** List examples of the ancient Greeks' many accomplishments.

_______________________________________________

_______________________________________________

**3. Analyze** Explain why northern Italy's economy is stronger than southern Italy's.

_______________________________________________

_______________________________________________

**4. Cause and Effect** Why did Spain and Portugal become poor in the 1800s?

_______________________________________________

_______________________________________________

# Southern Europe

**Section 1**

**MAIN IDEAS**

1. Southern Europe's physical features include rugged mountains and narrow coastal plains.
2. The region's climate and resources support such industries as agriculture, fishing, and tourism.

## Key Terms and Places

**Mediterranean Sea**  sea that borders Southern Europe

**Pyrenees**  mountain range separating Spain and France

**Apennines**  mountain range running along the whole Italian Peninsula

**Alps**  Europe's highest mountains, located in northern Italy

**Mediterranean climate**  type of climate found across Southern Europe, with warm, sunny days and mild nights for most of the year

## Section Summary

### PHYSICAL FEATURES

Southern Europe is composed of three peninsulas—the Iberian, the Italian, and the Balkan—and some large islands. All of the peninsulas have coastlines on the **Mediterranean Sea**.

These peninsulas are largely covered with rugged mountains. The **Pyrenees** form a boundary between Spain and France. The **Apennines** run along the Italian Peninsula. The **Alps**—Europe's highest mountains—are in northern Italy. The Pindus Mountains cover much of Greece. The region also has coastal plains and river valleys, where most of the farming is done and where most of the people live. Crete, which is south of Greece, and Sicily, at the southern tip of Italy, are two of the larger islands in the region.

In addition to the Mediterranean Sea, the Adriatic, Aegean, and Ionian seas are important to Southern Europe. They give the people food and an easy way to travel around the region. The Po and the Tagus are two important rivers in Southern

> **What are the three peninsulas of Southern Europe?**
>
> ______________________________
>
> ______________________________

> **Circle the four mountain ranges in Southern Europe.**

**Section 1, *continued***

Europe. The Po flows across northern Italy. The Tagus, the region's longest river, flows across the Iberian Peninsula.

## CLIMATE AND RESOURCES

The climate in Southern Europe is called a **Mediterranean climate**. The climate is warm and sunny in the summer and mild and rainy in the winter. Southern Europe's climate is one of its most valuable resources. It supports the growing of many crops, and it attracts tourists.

The seas are another important resource in Southern Europe. Many of the region's cities are ports, shipping goods all over the world. In addition, the seas support profitable fishing industries.

## CHALLENGE ACTIVITY

**Critical Thinking: Analyzing** Explain how Southern Europe's climate supports the region's economy.

> **What are the characteristics of a Mediterranean climate?**
> _______________________________
> _______________________________

# Southern Europe

**Section 2**

**MAIN IDEAS**

1. Early in its history, Greece was the home of a great civilization, but it was later ruled by foreign powers.
2. The Greek language, the Orthodox Church, and varied customs have helped shape Greece's culture.
3. In Greece today, many people are looking for new economic opportunities.

## Key Terms and Places

**Orthodox Church** a branch of Christianity that dates to the Byzantine Empire

**Athens** Greece's capital and largest city

## Section Summary

### HISTORY

Greece has been called the birthplace of Western culture. Ancient Greeks created statues and paintings that inspired later artists. They invented new forms of history and drama. They made advances in geometry. They developed a system of reason that is the basis for modern science, and they created democracy.

**Name two achievements of ancient Greece.**

In the 300s BC, Greece was conquered by Alexander the Great. Later, it was ruled by the Romans, the Byzantines, and the Ottoman Turks. Many Greeks were not happy with Turkish rule. In the 1800s they rebelled against the Turks and drove them out. Greece then became a monarchy.

**Circle the groups who ruled Greece before it became a monarchy.**

Greece experienced instability for much of the 1900s. A military dictatorship ruled from 1967 to 1974. More recently Greece has become democratic again.

### CULTURE

The Greek people today speak a form of the same language as their ancient ancestors, but their ancient religions have disappeared. Nearly everyone in Greece is a member of the **Orthodox Church**, a branch of

**To what church do most people in Greece belong?**





Interactive Reader and Study Guide

**Section 2,** *continued*

Christianity that dates to the Byzantine Empire. Religion is important to the Greek people and religious holidays are popular times for celebration.

The customs of the Greeks have been influenced by the country's geography and by past foreign rulers. Their customs rely on food native to Greece, and they have borrowed ideas for ingredients and food preparation from the groups who have ruled them. For centuries, the family has been central to Greek culture, and today it remains the cornerstone of Greek society.

> **What is the cornerstone of Greek society?**
> ____________________________

## GREECE TODAY

About three-fifths of all people in Greece live in cities. **Athens** is the largest city in Greece and its capital. Athens has both modern skyscrapers and ancient ruins. It is a major center of industry, and, as a result, suffers from air pollution.

People in the rural areas live much like they did in the past. Many live in isolated mountain villages. They grow crops, raise sheep and goats, and socialize in the village square.

Today, Greece's economy is growing but still lags behind other European nations. Greece has few mineral resources, and only one-fifth of the land is suitable for growing crops. However, Greece is a leading country in the shipping industry, with one of the largest shipping fleets in the world.

Another profitable industry is tourism. Many people from around the world are attracted to the ancient ruins of Athens and to the sandy beaches of Greece's islands.

> **Underline two important industries in Greece.**

## CHALLENGE ACTIVITY

**Critical Thinking: Synthesizing** Explain how religion and family work together to form a central part of Greek culture.

**Section 3**

---

**MAIN IDEAS**

**1.** Italian history can be divided into three periods: ancient Rome, the Renaissance, and unified Italy.

**2.** Religion and local traditions have helped shape Italy's culture.

**3.** Italy today has two distinct economic regions—northern Italy and southern Italy.

---

# Key Terms and Places

**pope**  the spiritual head of the Roman Catholic Church

**Vatican City**  an independent state within the city of Rome

**Sicily**  an island at Italy's southern tip

**Naples**  largest city in southern Italy and an important port

**Milan**  major industrial city in northern Italy and a fashion center

**Rome**  the capital of Italy

# Section Summary

### HISTORY

Ancient Rome grew from a tiny village in 700 BC to an empire that stretched from Britain in the northwest to the Persian Gulf. Ancient Rome's achievements in art, architecture, literature, law, and government still influence the world today. When the Roman Empire collapsed in the AD 400s, cities in Italy formed their own states. Later they became centers of trade. Merchants became rich and started supporting artists. The merchants' support of the arts helped lead to a period of great creativity in Europe called the Renaissance. During this time, artists and writers created some of the greatest works of art and literature in the world.

> **Underline important contributions of the Romans.**

> **What helped lead to the Renaissance?**
> ______________________________
> ______________________________

Italy remained divided until the mid-1800s when a rise in nationalism, or strong patriotic feelings for a country, led to a fight for unification. Italy was officially unified in 1861. In the 1920s Italy became a dictatorship under Benito Mussolini. This lasted until Italy's defeat in World War II, after which Italy became a democracy.

**Section 3,** *continued*

## CULTURE

The Roman Catholic Church has historically been the strongest influence on Italian culture. The **pope**, who is the spiritual head of the Roman Catholic Church, lives in **Vatican City**, which is an independent state located within Rome. Religious holidays and festivals are major events.

Local traditions have also influenced Italian culture. For example, variations in Italian food are based on local preferences and products. Other traditions reflect Italy's past. Italians have long been trendsetters in contemporary art forms, including painting, composing, fashion, and film.

> Underline three influences on Italian culture.

## ITALY TODAY

Southern Italy is poorer than northern Italy. It has less industry and relies on agriculture and tourism for its survival. Farming is especially important in **Sicily**, an island at the southern tip of Italy. **Naples** is the largest city in southern Italy and a busy port.

Northern Italy has a strong economy, including major industrial centers, the most productive farm-land, and the most popular tourist destinations. **Milan** is a major industrial center and a worldwide center for fashion design. Turin and Genoa are also industrial centers. Florence, Pisa, and Venice are popular tourist destinations. **Rome**, Italy's capital, is located between northern Italy and southern Italy.

> Why is the economy of northern Italy strong?
>
> _______________________
>
> _______________________

## CHALLENGE ACTIVITY

**Critical Thinking: Analyzing and Evaluating** Identify three important periods in Italy's history, and explain why each one is important to Italy today.

# Southern Europe

## Section 4

**MAIN IDEAS**

1. Throughout their histories, Spain and Portugal have been part of many large and powerful empires.
2. The cultures of Spain and Portugal reflect their long histories.
3. Having been both rich and poor in the past, Spain and Portugal today have growing economies.

## Key Terms and Places

**Iberia**  westernmost peninsula in Europe

**parliamentary monarchy**  form of government in which a king rules with the help of an elected parliament

**Madrid**  capital of Spain

**Barcelona**  center of industry, culture, and tourism in Spain

**Lisbon**  large city in Portugal and important industrial center

## Section Summary

### HISTORY

Spain and Portugal lie on the Iberian Peninsula, or **Iberia**. Both countries have been part of large, powerful empires. Coastal areas of what is now Spain were first ruled by Phoenicians from the eastern Mediterranean. Later the Greeks established colonies there. A few centuries later, Iberia became part of the Roman Empire. After the fall of Rome, they were conquered by the Moors—Muslims from north Africa. For about 600 years, much of the Iberian Peninsula was under Muslim rule.

Eventually the Christian kingdoms of Spain and Portugal banded together to drive out the Muslims. Both countries went on to establish empires of their own in the Americas, Africa, and Asia. Both countries became rich and powerful until the 1800s, when most of their colonies broke away and became independent.

> **What foreign powers have ruled Spain and Portugal?**
> ________________________
> ________________________

> **Where did Spain and Portugal establish colonies?**
> ________________________
> ________________________



  Interactive Reader and Study Guide

Section 4, *continued*

## CULTURE

Many dialects of Spanish and Portuguese are spoken in various parts of Iberia. In addition, Catalan, which is similar to Spanish, is spoken in eastern Spain. Galician, which is more closely related to Portuguese, is spoken in northwest Spain. The Basques of the Pyrenees have their own language and customs, unlike those of the rest of Spain. In both Spain and Portugal the people are mainly Roman Catholic.

Music is important to both countries. The Portuguese are famous for sad folk songs called fados. The Spanish are famous for a style of song and dance called flamenco. Much of the peninsula's art and architecture reflect its Muslim past. The round arches and elaborate tilework on many buildings were influenced by Muslim design.

> **Circle the group in Spain that has its own language and customs.**

> **What Muslim influences can be seen on buildings in Iberia?**
> ______________________________
> ______________________________

## SPAIN AND PORTUGAL TODAY

With the gold and silver found in their colonies, Spain and Portugal were once the richest countries in Europe. When other countries started building industrial economies, Spain and Portugal continued to rely on their colonies. When their colonies broke away, the income they had depended on was lost. Today the economies of Spain and Portugal are growing rapidly, but they remain poorer than many countries in Europe.

Spain is governed by a **parliamentary monarchy**—a king rules with the help of an elected parliament. **Madrid**—the capital—and **Barcelona** are centers of industry, culture, and tourism.

Portugal is a republic whose leaders are elected. The economy is based on industries in **Lisbon** and other large cities. In rural areas farmers grow many crops but are most famous for grapes and cork.

> **Underline the definition of a parliamentary monarchy.**

## CHALLENGE ACTIVITY

**Critical Thinking: Comparing and Contrasting** Compare and contrast Spain and Portugal. Then explain which country you would choose to visit first and why.

# West-Central Europe

## CHAPTER SUMMARY

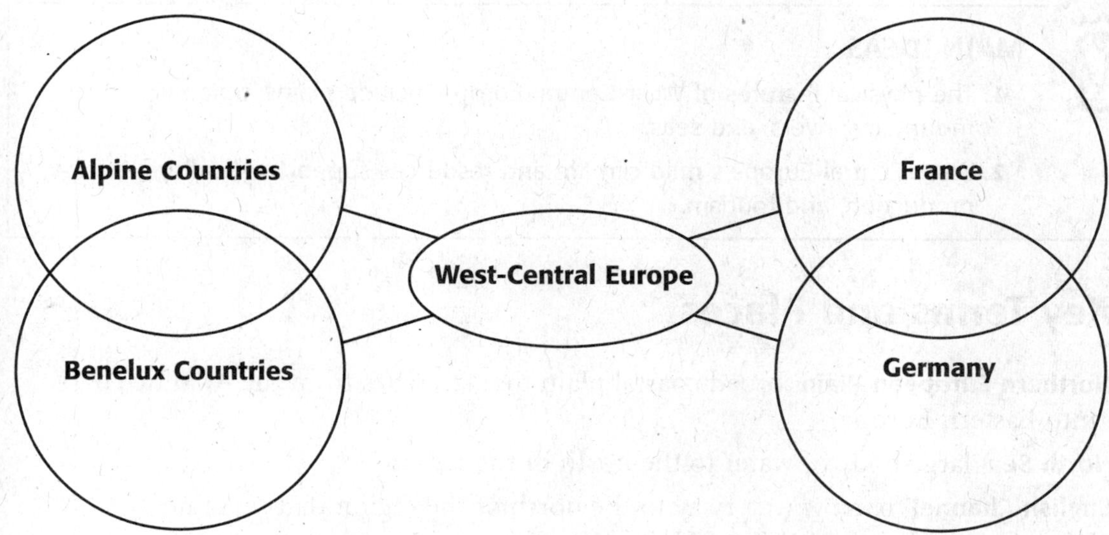


## COMPREHENSION AND CRITICAL THINKING

Complete the graphic organizer by filling in the areas shared by two
countries or regions with features they share in common. Use
information from the graphic organizer to compare and contrast
the countries of the region.

1. **Describe**  In what ways are Germany and France alike in their size, cultural
   achievements, and influence in the region?

   _______________________________________________________________

   _______________________________________________________________

2. **Explain**  What features of their physical geography have helped both the Benelux
   and Alpine countries build strong economies?

   _______________________________________________________________

   _______________________________________________________________

3. **Understanding Cause and Effect**  How are the histories of Germany and France
   in the twentieth century closely connected?

   _______________________________________________________________

   _______________________________________________________________

## West-Central Europe

**Section 1**

**MAIN IDEAS**

1. The physical features of West-Central Europe include plains, uplands, mountains, rivers, and seas.

2. West-Central Europe's mild climate and resources support agriculture, energy production, and tourism.

# Key Terms and Places

**Northern European Plain** broad coastal plain that stretches from the Atlantic coast into Eastern Europe

**North Sea** large body of water to the north of the region

**English Channel** narrow waterway to the north of the region that separates West-Central Europe from the United Kingdom

**Danube River** one of the major rivers of the region

**Rhine River** one of the major rivers of the region

**navigable river** river that is deep and wide enough for ships to use

# Section Summary

### PHYSICAL FEATURES

West-Central Europe has three major types of landforms: plains, uplands, and mountains. Most of the **Northern European Plain** is flat or rolling, but in the Netherlands the plains drop below sea level. The plain has the region's best farmland and largest cities. The Central Uplands are in the middle of the region. This area has many rounded hills, small plateaus, and valleys. In France, the uplands include the Massif Central, a plateau region, and the Jura Mountains. Coal fields in the Central Uplands have helped to make it a major mining and industrial area. The area has some fertile soil, but is mostly too rocky for farming.

The region has two high mountain ranges. The Alps and Pyrenees form the alpine mountain system. The Alps are the highest mountains in Europe.

> Circle the three major landform types in West-Central Europe.

> Underline the names of three mountain ranges in West-Central Europe.

Water is an important part of the region's physical geography. The Mediterranean Sea borders France to the south. The Atlantic Ocean lies to the west and the **North Sea** and the **English Channel** lie to the north. The **Danube** and the **Rhine** rivers are important waterways for trade and travel. Several of the region's rivers are **navigable**. These rivers and a system of canals link the region's interior to the seas.

> Circle two important rivers in the region.

## CLIMATE AND RESOURCES

Most of West-Central Europe has a marine west coast climate. This is a mild climate with colder winters. In the Alps and other higher elevation areas, the climate is colder and wetter. In contrast, southern France has a warm Mediterranean climate with dry, hot summers and mild, wet winters.

The mild climate is a valuable resource. Mild temperatures, ample rainfall, and rich soil have made the region's farmlands very productive. Farmers grow grapes, grains, and vegetables. In the Alps and the uplands, farmers raise livestock.

> Why is a mild climate a valuable resource for the region?
>
> _______________________
> _______________________

Energy resources are not evenly divided. France has iron ore and coal. Germany has coal, and the Netherlands has natural gas. Fast-flowing alpine rivers provide hydroelectric power. Even so, many countries have to import fuel. The Alps are another important resource. Tourists come to the mountains for the scenery and to ski and hike.

> Circle the energy resources of France. Underline the energy resources of Germany and the Netherlands.

> In what way are the Alps an important resource for the region?
>
> _______________________

## CHALLENGE ACTIVITY

**Critical Thinking: Evaluating Information** How have landforms and bodies of water affected activities in the region? Give support for your answer.

# West-Central Europe

## Section 2

**MAIN IDEAS**

1. During its history France has been a kingdom, empire, colonial power, and republic.
2. The culture of France has contributed to the world's arts and ideas.
3. France today is a farming and manufacturing center.
4. The Benelux Countries have strong economies and high standards of living.

## Key Terms and Places

**Paris**  capital and largest city in France

**Amsterdam**  capital of the Netherlands

**The Hague**  seat of government in the Netherlands

**Brussels**  capital of Belgium, headquarters of many international organizations

**cosmopolitan**  characterized by many foreign influences

## Section Summary

### HISTORY OF FRANCE

France has a history as a major European power. Early in its history, the Romans and Franks conquered lands that became France. The Franks' ruler, Charlemagne, built a large empire. Later the Normans claimed northwestern France. They extended their rule to England, becoming kings of England and ruling part of France as well. English rule in France ended in 1453 after the Hundred Years' War. In the 1500s France became a colonial power with colonies in Asia, Africa, and the Americas.

In 1789 the French Revolution led to greater equality and more rights for the French. In 1799 Napoleon Bonaparte took power. He conquered most of Europe, creating a vast empire, which ended in 1815 when European powers defeated his armies. In the 1900s during World Wars I and II, Germany invaded France. In the 1950s and 1960s, many French colonies declared independence. France is now a democratic republic.

> Circle the names of three groups that ruled France during its early history.

> Underline the sentence that tells how Napoleon Bonaparte affected French history.

## THE CULTURE OF FRANCE

The French share a common heritage. Most speak French and are Catholic. Recently France has become more diverse due to immigrants. The French share a love of good food and company. The French have made major contributions to the arts and ideas—including impressionism, Gothic cathedrals, and Enlightenment ideas about government.

> Underline the sentences that describe the ways the French are alike.

## FRANCE TODAY

France is a European and world leader. **Paris** is a center of business, finance, learning, and culture. France has a strong economy and it is the EU's top agricultural producer. Its economy also relies on tourism and the export of goods such as perfumes and wines.

## THE BENELUX COUNTRIES

Belgium, the Netherlands, and Luxembourg are the Benelux Countries. Their location has led to invasions but has also promoted trade. All are densely populated, lie at low elevations between larger, stronger countries, and have strong economies and democratic governments. North Sea harbors have made the Netherlands a center for trade. Major cities include Rotterdam, **Amsterdam**, and **The Hague**. In Belgium, **Brussels** is a **cosmopolitan** city with many international organizations. Luxembourg's economy is based on banking and steel and chemical production.

> Circle the names of the three Benelux Countries.

> Underline the sentences that tell how the Benelux Countries are alike.

## CHALLENGE ACTIVITY

**Critical Thinking: Drawing Conclusions** How has the location of the Benelux Countries both helped and hurt them?

# West-Central Europe

**Section 3**

> **MAIN IDEAS**
>
> 1. After a history of division and two world wars, Germany is now a unified country.
> 2. German culture, known for its contributions to music, literature, and science, is growing more diverse.
> 3. Germany today has Europe's largest economy, but eastern Germany faces challenges.
> 4. The Alpine Countries reflect German culture and have strong economies based on tourism and services.

## Key Terms and Places

**Berlin** capital of Germany that divided East Germany and West Germany

**chancellor** prime minister elected by Parliament who runs the government

**Vienna** Austria's capital and largest city

**cantons** districts of Switzerland's federal republic

**Bern** capital of Switzerland

## Section Summary

### HISTORY OF GERMANY

The land that is now Germany was a loose association of small states for hundreds of years. In 1871 Prussia united them to create Germany, which grew into a world power. In the 1900s Germany lost two world wars. After its defeat in World War II, Germany was divided into West Germany and East Germany. West Germany kept control of western **Berlin** and Communists took eastern Berlin. They built the Berlin Wall to stop East Germans from escaping. In 1989 democracy movements swept Eastern Europe and communism collapsed. In 1990 East and West Germany were reunited.

> **Why did Communist leaders build the Berlin Wall?**
>
> _______________________________
>
> _______________________________

> **What events helped East and West Germany reunite?**
>
> _______________________________
>
> _______________________________

### CULTURE OF GERMANY

Most people in Germany are ethnic Germans who speak German. Germany also has immigrants from Turkey, Italy, and Eastern Europe. Most people

## Section 3, *continued*

are either Protestant or Catholic. In the formerly Communist East Germany some have no religious ties. Germans have made important contributions to classical music, literature, chemistry, engineering, medicine, and physics.

**In what ways do Germans differ in their religious beliefs?**

_______________________________

_______________________________

## GERMANY TODAY

Today Germany is a leading European power. A parliament with a **chancellor** governs the nation. Berlin is the largest city. Germany is Europe's largest economy, exporting many products, including cars. Its economy is based mainly on industries such as chemicals, engineering, and steel, but agriculture is also still important. The reuniting of East and West Germany slowed economic growth. East Germany's economy continues to lag.

**What economic activities have made Germany Europe's strongest economy?**

_______________________________

_______________________________

## THE ALPINE COUNTRIES

Austria and Switzerland are the Alpine Countries. Both were once part of the Holy Roman Empire, are landlocked, influenced by German culture, and prosperous. Austria once led the powerful Habsburg Empire that ruled much of Europe. After World War I the empire collapsed. Today Austria is a modern, industrialized nation. **Vienna** is a center of music and fine arts. Austria has a strong economy with little unemployment. Service industries and tourism are important.

**What do the Alpine Countries have in common?**

_______________________________

_______________________________

**Circle important industries in the Alpine Countries.**

Switzerland has been independent since the 1600s. It is a federal republic made up of 26 **cantons**. It is neutral and not a member of the EU or NATO. The Swiss speak several languages, including German and French. Its capital is **Bern**. The country is known for its banks, watches and other precision devices, chocolate, and cheese.

## CHALLENGE ACTIVITY

**Critical Thinking: Drawing Inferences** How have the Alpine Countries used their human resources to become prosperous? Give support for your answer.



 Interactive Reader and Study Guide

# Northern Europe

## CHAPTER SUMMARY

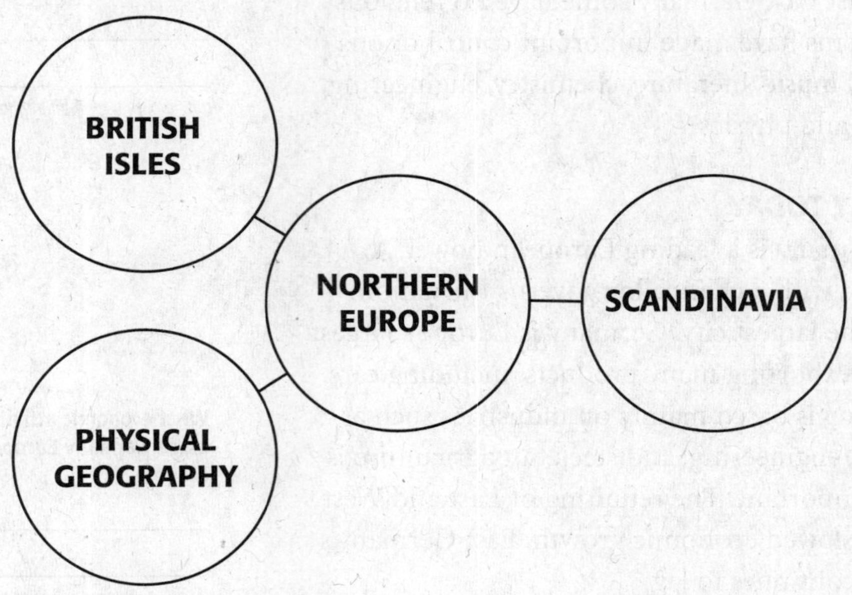


## COMPREHENSION AND CRITICAL THINKING

Use the graphic organizer to answer the following questions.

**1. Identify** Which circle would you use to list details about the natural resources of Northern Europe?

_______________________________________________________________

**2. Compare** How are the climates of the British Isles and Scandinavia similar?

_______________________________________________________________

_______________________________________________________________

**3. Analyze** Why would details about history fit into more than one circle above?

_______________________________________________________________

_______________________________________________________________

**4. Make a Generalization** If details about economy were added to the "British Isles" and "Scandinavia" circles, how could you summarize the details in a single statement?

_______________________________________________________________

_______________________________________________________________

# Northern Europe

**Section 1**

**MAIN IDEAS**

1. The physical features of Northern Europe include low mountain ranges and jagged coastlines.
2. Northern Europe's natural resources include energy sources, soils, and seas.
3. The climates of Northern Europe range from a mild coastal climate to a freezing ice cap climate.

## Key Terms and Places

**British Isles** a group of islands located across the English Channel from the rest of Europe

**Scandinavia** a region of islands and peninsulas in far northern Europe

**fjord** a narrow inlet of the sea set between high, rocky cliffs

**geothermal energy** energy from the heat of Earth's interior

**North Atlantic Drift** an ocean current that brings warm, moist air across the Atlantic Ocean

## Section Summary

### PHYSICAL FEATURES

Northern Europe consists of two regions. The **British Isles** are a group of islands located across the English Channel from the rest of Europe. **Scandinavia** is a region of islands and peninsulas in far northern Europe. Iceland, to the west, is often considered part of Scandinavia.

Fewer people live in the northern portion of the region, which is covered by rocky hills and low mountains. Farmland and plains stretch across the southern part of the region.

Slow moving sheets of ice called glaciers once covered the region. They carved lakes and **fjords**, narrow inlets between high, rocky cliffs. The fjords make the coast of Norway irregular and jagged.

> Underline the sentence that describes the land in the northern portion of the region.

> What two features were created by glaciers?
>
> _______________________

### NATURAL RESOURCES

Northern Europe has many natural resources that have helped make it one of the world's wealthiest



 Interactive Reader and Study Guide

## Section 1, *continued*

regions. Energy resources include oil and natural gas in areas of the North Sea controlled by the United Kingdom and Norway. Hydroelectric energy is created by lakes and rivers. Iceland's hot springs produce **geothermal energy**, or energy from the heat of the Earth's interior.

Forests in Norway, Sweden, and Finland provide timber. Fertile farmland in southern areas provides crops such as wheat and potatoes. The seas and oceans that surround the region have provided fish to the people of Northern Europe for centuries.

> **Underline the sentences that describe energy resources located in the North Sea.**

## CLIMATES

Although much of the region is very far north and close to the Arctic Circle, the climates in Northern Europe are surprisingly mild. The **North Atlantic Drift** is a warm ocean current that brings warm, moist air across the Atlantic Ocean to Northern Europe. It creates warmer temperatures than other areas located as far north.

Much of the region has a marine west coast climate with mild summers and frequent rainfall. Central Norway, Sweden, and southern Finland have a humid continental climate with four seasons. Farther north are subarctic regions, with long, cold winters and short summers. Tundra and ice cap climates produce extremely cold temperatures year-round.

> **How does the North Atlantic Drift affect the climate?**
>
> _______________________________
>
> _______________________________

## CHALLENGE ACTIVITY

**Critical Thinking: Making Inferences** Fewer people live in the northern portion of the region than in the southern portion. List all of the factors that you can think of which might help explain this pattern.





Interactive Reader and Study Guide

## Northern Europe

Section 2

### MAIN IDEAS

1. Invaders and a global empire have shaped the history of the British Isles.
2. British culture, such as government and music, has influenced much of the world.
3. Efforts to bring peace to Northern Ireland and maintain strong economies are important issues in the British Isles today.

# Key Terms and Places

**constitutional monarchy** a type of democracy in which a king or queen serves as head of state, but a legislature makes the laws

**Magna Carta** a document that limited the powers of kings and required everyone to obey the law

**disarm** give up all weapons

**London** the capital of the United Kingdom

**Dublin** the capital of Ireland

# Section Summary

### HISTORY

The Republic of Ireland and the United Kingdom make up the British Isles. The United Kingdom consists of England, Scotland, Wales, and Northern Ireland.

> **What two countries make up the British Isles?**
>
> ___________________

Celts, Romans, Angles, Saxons, Vikings, and Normans invaded Britain in its early history. Over time England grew in strength, and by the 1500s it had become a world power. England eventually formed the United Kingdom with Wales, Scotland, and Ireland. It then developed a strong economy thanks to the Industrial Revolution and its colonies abroad.

> **Underline the sentence that lists the countries that formed the United Kingdom.**

The British Empire stretched around the world by 1900, but later declined. The Republic of Ireland won its independence in 1921. By the mid-1900s Britain had given up most of its colonies.

## CULTURE

The United Kingdom is a **constitutional monarchy**, a type of democracy in which a monarch serves as head of state, but a legislature makes the laws. England first limited the power of monarchs during the Middle Ages in a document called **Magna Carta**. Ireland has a president that serves as head of state and a prime minister who runs the government along with Parliament.

The people of the British Isles share many culture traits, but each culture is also unique. The people of Ireland and Scotland keep many traditions alive, and immigrants from all over the world add new traits to the culture of the British Isles.

British popular culture has influenced people around the world. British literature and music are well known and the English language is used in many countries.

> In what ways are the governments of the United Kingdom and Ireland similar and different?
> ____________________
> ____________________
> ____________________

> Underline the sentence that describes the importance of British popular culture.

## BRITISH ISLES TODAY

Conflict in Northern Ireland remains an important issue. Many Catholics there feel they have not been treated fairly by Protestants who control the government. Some groups have refused to **disarm**, or give up all weapons. But hopes are high that a peaceful settlement will be reached.

The economies in the British Isles are very strong. **London**, the capital of the United Kingdom, is a center for world trade, and the country has reserves of oil and natural gas in the North Sea. **Dublin**, the capital of Ireland, has attracted new industries like computers and electronics.

> What two groups are in conflict in Northern Ireland?
> ____________________
> ____________________

## CHALLENGE ACTIVITY

**Critical Thinking: Identify Cause and Effect** Make a list of well-known British authors and/or musicians. Select one for further research and write a brief report about his or her accomplishments and influence.

## Northern Europe

### Section 3

**MAIN IDEAS**

1. The history of Scandinavia dates back to the time of the Vikings.
2. Scandinavia today is known for its peaceful and prosperous countries.

## Key Terms and Places

**Vikings**  Scandinavian warriors who raided Europe in the early Middle Ages

**Stockholm**  Sweden's capital and largest city

**neutral**  not taking sides in an international conflict

**uninhabitable**  not able to support human settlement

**Oslo**  the capital of Norway

**Helsinki**  Finland's capital and largest city

**geyser**  a spring that shoots hot water and steam into the air

## Section Summary

### HISTORY

**Vikings** were Scandinavian warriors who raided Europe during the Middle Ages. They were greatly feared and conquered the British Isles, Finland, and parts of France, Germany, and Russia.

Vikings were excellent sailors. They were the first Europeans with settlements in Iceland and Greenland and the first Europeans to reach North America. They stopped raiding in the 1100s and focused on strengthening their kingdoms. Norway, Sweden, and Denmark competed for control of the region, and by the late 1300s Denmark ruled all the Scandinavian Kingdoms and territories. Sweden eventually broke away, taking Finland, and later Norway, with it. Norway, Finland, and Iceland became independent countries during the 1900s. Greenland remains part of Denmark as a self-ruling territory.

> Underline the sentence that tells areas that Vikings explored.

> What three countries did not become independent until the 1900s?
>
> ___________________________
>
> ___________________________

## SCANDINAVIA TODAY

Scandinavians share many things, including culture traits like similar political views, languages, and religions. They enjoy high standards of living, are well-educated, and get free health care. The countries have strong economies and large cities.

Each country is unique as well. Sweden has the largest area and population. Most people live in the south in large towns and cities. **Stockholm** is Sweden's capital and largest city. For about 200 years, Sweden has been **neutral**, choosing not to take sides in international conflicts.

Denmark is the smallest country and the most densely populated. Its economy relies on excellent farmland and modern industries. Greenland is a territory of Denmark but most of it is covered with ice and is **uninhabitable**, or not able to support human settlement.

Norway has one of the longest coastlines in the world. **Oslo** is Norway's capital, a leading seaport and industrial center. Important industries include timber, shipping, and fishing. Oil and natural gas provide Scandinavia with the highest per capita GDP in the region; however, oil fields in the North Sea are expected to run out during the next century.

Finland relies on trade and exports paper and other forest products. Shipbuilding and electronics are also important. **Helsinki** is its capital and largest city.

Iceland has fertile farmland and rich fishing grounds. Tourists come to see its volcanoes, glaciers, and **geysers**, springs that shoot hot water and steam into the air. Geothermal energy heats many buildings.

> Underline the sentence that lists culture traits Scandinavians share.

> What is the largest country in area? What is the smallest?
> _______________________
> _______________________

> Why are tourists attracted to Iceland?
> _______________________
> _______________________

## CHALLENGE ACTIVITY

**Critical Thinking: Compare and Contrast** Compare and contrast the countries of Scandinavia with the United States. In what ways are they similar? In what ways are they different?

# Eastern Europe

## CHAPTER SUMMARY

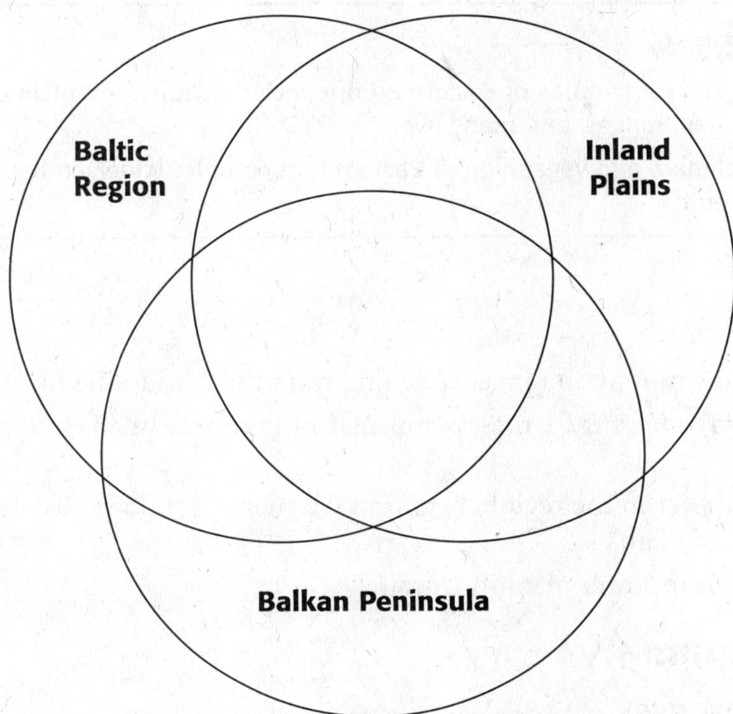


## COMPREHENSION AND CRITICAL THINKING

Complete the outer areas of the graphic organizer by describing unique features of each region of Eastern Europe. Complete the center areas where the circles meet by describing what the regions of Eastern Europe have in common. Use information from the graphic organizer to compare and contrast the countries of the region.

**1. Describe** In what ways are the cultures of Eastern Europe very diverse?

_______________________________________________

_______________________________________________

**2. Explain** What features of physical geography have made the three areas of Eastern Europe different from each other?

_______________________________________________

_______________________________________________

**3. Analyze Information** Which area has faced the most serious problems since the end of Communist rule? Give reasons for your answer.

_______________________________________________

_______________________________________________

# Eastern Europe

**Section 1**

**MAIN IDEAS**

1. The physical features of Eastern Europe include wide open plains, rugged mountain ranges, and many rivers.
2. The climate and vegetation of Eastern Europe differ widely in the north and the south.

## Key Places

**Carpathians** a low mountain range stretching from the Alps to the Black Sea area

**Balkan Peninsula** one of the largest peninsulas in Europe, extends into the Mediterranean

**Danube** longest river in the region, begins in Germany and flows east across the Great Hungarian Plain

**Chernobyl** a nuclear power plant in Ukraine

## Section Summary

### PHYSICAL FEATURES

The landforms of Eastern Europe stretch across the region in broad bands of plains and mountains. The Northern European Plain covers most of northern Europe. The **Carpathians** are south of the plains. These mountains extend from the Alps to the Black Sea area. South of the Carpathians is another large plain, the Great Hungarian Plain, located mostly in Hungary. South of this plain are the Dinaric Alps and Balkan Mountains. These mountain ranges cover most of the **Balkan Peninsula**. The peninsula extends into the Mediterranean.

> Circle three mountain ranges in Eastern Europe.

Eastern Europe has many water bodies that are important routes for transportation and trade. The Adriatic Sea lies to the southwest. The Black Sea is east of the region. The Baltic Sea is in the far north. It remains frozen some part of the year, reducing its usefulness.

The rivers that flow through Eastern Europe are also important for trade and transportation, especially the **Danube**. The Danube crosses nine

> Why are rivers important to the economy of Eastern Europe?
> _______________________

## Section 1, *continued*

countries before it empties into the Black Sea. This river is very important to Eastern Europe's economy. Many of the region's largest cities are along its banks. Dams on the river provide electricity for the region. This busy river has become very polluted from heavy use.

> **Why is the Danube so polluted?**
> ____________________________

## CLIMATE AND VEGETATION

Types of climates and vegetation in Eastern Europe vary widely. The shores of the Baltic Sea in the far north have the coldest climate. The area does not get much rain, but is sometimes foggy. Its cold, damp climate allows huge forests to grow.

The interior plains have a much milder climate than the Baltic Region. Winters can be very cold, but summers are mild. The western parts of the interior plains get more rain than the eastern parts. Because of its varied climate, the forests cover much of the north and open grassy plains lie in the south. In 1986 Eastern Europe's forests were damaged by a major nuclear accident at **Chernobyl**. An explosion released huge amounts of radiation into the air that poisoned forests and ruined soil across the region.

The Balkan coast has a Mediterranean climate with warm summers and mild winters. Some of its beaches attract tourists. The area does not get much rain so there are not many forests. The land is covered by shrubs and trees that do not need much water.

> **Circle the type of climate found in the Balkan Peninsula.**

> **Why is the vegetation of the Balkan Peninsula different from the vegetation of the Baltic Region?**
> ____________________________
> ____________________________

## CHALLENGE ACTIVITY

**Critical Thinking: Analyzing Information** How has climate affected the vegetation in Eastern Europe? Explain your answer in a brief paragraph.

# Eastern Europe

## Section 2

**MAIN IDEAS**

1. History ties Poland and the Baltic Republics together.
2. The cultures of Poland and the Baltic Republics differ in language and religion but share common customs.
3. Economic growth is a major issue in the region today.

## Key Terms and Places

**infrastructure** the set of resources—like roads, airports, and factories—that a country needs in order to support economic activities

**Warsaw** capital of Poland

## Section Summary

### HISTORY

The groups who settled around the Baltic Sea in ancient times developed into the Estonians, Latvians, Lithuanians, and Polish. Each group had its own language and culture, but in time they became connected by a shared history. By the Middle Ages, each had formed an independent kingdom. Lithuania and Poland were the largest and strongest. They ruled large parts of Eastern and Northern Europe. Latvia and Estonia were smaller, weaker, and often invaded.

In the 1900s two world wars greatly damaged the Baltic region. In World War I millions died in Poland, and thousands more were killed in the Baltic countries. The region also suffered greatly during World War II. This war began with the invasion of Poland by Germany. While fighting the Germans, troops from the Soviet Union invaded Poland and occupied Estonia, Latvia, and Lithuania.

After World War II ended, the Soviet Union took over much of Eastern Europe, making Estonia, Latvia, and Lithuania part of the Soviet Union. The Soviets also forced Poland to accept a Communist

> **How did the Baltic Republics lose their independence after World War II?**
>
> ____________________________





Interactive Reader and Study Guide

## Section 2, continued

government. In 1989 Poland rejected communism and elected new leaders. In 1991 the Baltic Republics broke away from the Soviet Union. They became independent countries again.

> **Underline the sentences that tell how the governments of the Baltic countries changed after 1989.**

## CULTURE

The Baltic countries differ from each other in languages and religion, but are alike in other ways. The Latvian and Lithuanian languages are alike, but Estonian is like Finnish, the language spoken in Finland. Polish is more like languages of countries farther south. Most Polish people and Lithuanians are Roman Catholics because their trading partners were from Catholic countries. Latvians and Estonians are Lutherans, because these countries were once ruled by Sweden, where most people are Lutherans.

These countries share many customs and practices. They eat similar foods, practice crafts such as ceramics, painting, and embroidery, and enjoy music and dance.

## THE REGION TODAY

The economies of Baltic countries grew more slowly than those of Western European nations because the Soviets did not build a decent **infrastructure**. Poland and the Baltic Republics are working hard to rebuild their economies. As a result, cities like **Warsaw** have become major industrial centers. To help their economies grow, many Baltic countries are trying to attract more tourists. Since the Soviet Union collapsed in 1991, the rich cultures, historic sites, and cool summer climates attract visitors.

> **What economic problems did Soviet rule cause for Poland and the Baltic Republics?**
>
> _______________________________
>
> _______________________________

> **What new source of income have the Baltic countries found?**
>
> _______________________________

## CHALLENGE ACTIVITY

**Critical Thinking: Understanding Effects** Imagine that you are a newly elected leader of a Baltic country. Write a speech that tells what you will do to improve the country's economy.



 Interactive Reader and Study Guide

# Eastern Europe

**Section 3**

**MAIN IDEAS**

1. The histories and cultures of inland Eastern Europe vary from country to country.

2. Most of inland Eastern Europe today has stable governments, strong economies, and influential cities.

## Key Terms and Places

**Prague**  capital of the Czech Republic

**Kiev**  present-day city where the Rus built a settlement that eventually became the center of a huge empire

**Commonwealth of Independent States**  CIS, an international alliance that meets to discuss issues such as trade and immigration that affect former Soviet republics

**Budapest**  capital of Hungary

## Section Summary

### HISTORY AND CULTURE

Inland Eastern Europe consists of the Czech Republic, Slovakia, Hungary, Ukraine, Belarus, and Moldova. It is located on the Northern European and Hungarian plains. The area of the Czech Republic and Slovakia was settled by Slavs, people from Asia who moved into Europe by 1000 BC. The area was once made up of many small Slavic kingdoms. In time more powerful countries like Austria conquered these kingdoms. After Austria was defeated in World War I, land was taken away from it to create the nation of Czechoslovakia. In 1993 Czechoslovakia split into the Czech Republic and Slovakia.

What group of people settled the area of the Czech Republic and Slovakia?

_______________________________

Circle the country that land was taken from to create Czechoslovakia.

When was Czechoslovakia created and when did it split?

_______________________________

_______________________________

Because the Czech Republic and Slovakia are located in Central Europe, they have long had ties to Western Europe. Many people are Roman Catholic. The architecture of **Prague** and other cities shows Western influences.

In the 900s Magyar people invaded what is now Hungary. The Magyars influenced Hungarian



  Interactive Reader and Study Guide

**Section 3, *continued***

culture, especially their language. It developed from
the Magyar language.

Ukraine, Belarus, and Moldova are more closely
tied to Russia than to the West. The Rus people
built the settlement that is now **Kiev**. The rulers
of Kiev created a huge empire that became part
of Russia in the late 1700s. In 1922 Russia became
the Soviet Union. Ukraine, Belarus, and Moldova
became Soviet republics. After the breakup of the
Soviet Union in 1991, these countries became
independent. Russia has strongly influenced their
cultures. Most people are Orthodox Christians,
and the Ukrainian and Belarusian languages are
written in the Cyrillic, or Russian, alphabet.

> Underline two ways the cultures of Ukraine, Belarus, and Moldova have been influenced by Russia.

## INLAND EASTERN EUROPE TODAY

All  inland Eastern European countries were once
either part of the Soviet Union or run by Soviet-
influenced Communist governments. People
had few freedoms. The Soviets did a poor job
of managing these economies. Since Soviet rule
ended, all of these countries have become republics,
although Belarus is run by a dictator.

> How did the government and the economy of the region change after the collapse of the Soviet Union?
>
> ________________________
>
> ________________________
>
> ________________________
>
> ________________________

Countries of the region belong to international
unions. Belarus, Ukraine, and Moldova belong to
the **Commonwealth of Independent States**. Its
members meet to talk about such issues as trade
and immigration. The Czech Republic, Slovakia,
and Hungary seek closer ties to the West and belong
to the EU. Since the collapse of the Soviet Union,
the Czech Republic, Slovakia, Hungary, and Ukraine
have become prosperous industrial centers.

The capital cities of the region are also economic
and cultural centers. Prague, Kiev, and **Budapest** are
especially important. They are the most prosperous
cities in the region. Many tourists visit these cities.

## CHALLENGE ACTIVITY

**Critical Thinking: Evaluating** Write a brief essay that explains how the
location of inland Eastern Europe has affected its culture and history.

# Eastern Europe

**Section 4**

**MAIN IDEAS**

1. The history of the Balkan countries is one of conquest and conflict.
2. The cultures of the Balkan countries are shaped by the many ethnic groups who live there.
3. Civil War and weak economies are major challenges to the region today.

## Key Terms

**ethnic cleansing** effort to remove all members of a group from a country or region

## Section Summary

### HISTORY

The Balkan Peninsula has been ruled by many different groups. In ancient times, the Greeks founded colonies near the Black Sea. This area is now Bulgaria and Romania. Next the Romans conquered most of the area between the Adriatic Sea and the Danube River. When the Roman Empire divided into west and east in AD 300s, the Balkan Peninsula became part of the Byzantine Empire. Under Byzantine rule many people became Orthodox Christians.

Over a thousand years later, Muslim Ottoman Turks conquered the Byzantine Empire. Many people in the Balkans became Muslims. Ottoman rule lasted until the 1800s, when the people of the Balkans drove the Ottomans out. They then created their own kingdoms. In the late 1800s, the Austria-Hungarian Empire took over part of the peninsula. To protest the takeover, a man from Serbia shot the heir to the Austro-Hungarian throne. This event led to World War I. After the war Europe's leaders combined many formerly independent countries into Yugoslavia. In the 1990s this country broke apart. Ethnic and religious conflict led to its collapse.

> Underline the names of groups or empires that ruled the Balkan Peninsula until the 1800s.





Interactive Reader and Study Guide

## CULTURE

The Balkans are Europe's most diverse region. People practice different religions and speak many different languages. Most people are Christians. They are either Orthodox, Roman Catholic, or Protestant. Islam is also practiced. In Albania most people are Muslims. Most people in the Balkans speak languages from one of three language groups. Some speak Slavic languages related to Russian. People in Romania speak Germanic languages and a language that comes from Latin.

> **What makes the Balkans Europe's most diverse region?**
>
> ___________________________
>
> ___________________________
>
> ___________________________

## THE BALKANS TODAY

Balkan countries were once run by Communist governments. Poor economic planning has hurt the economies of the region. It is still the poorest region in Europe today. After the breakup of Yugoslavia, many of the countries that had been a part of this country faced violent ethnic and religious conflicts. These conflicts led to **ethnic cleansing** in some areas. In 1995 troops from all over the world came to Bosnia and Herzegovina to help end the fighting.

> **What caused the violence in the Balkans after Yugoslavia broke apart? What stopped the fighting?**
>
> ___________________________
>
> ___________________________
>
> ___________________________

The Balkan region now has eight countries. Five of them were once part of Yugoslavia. Macedonia was the only country to break away peacefully. When Croatia broke away, fighting started between ethnic Croats and Serbs. Peace did not return until many Serbs left Croatia. The two countries of Bosnia and Herzegovina and Serbia and Montenegro also had terrible ethnic fighting.

The Balkans includes three other countries. Albania and Romania are poor and have serious economic and political problems. Since the fall of the Soviet Union, Bulgaria has built a strong market economy based on industry and tourism.

## CHALLENGE ACTIVITY

**Critical Thinking: Predicting** Write a paragraph that explains how ethnic diversity could be an advantage for the Balkan countries in the future.

Name _________________________________ Class _______________ Date _____________

# Russia and the Caucasus

## CHAPTER SUMMARY

|  | People/ Culture | History | Government | Economy/ Resources |
|---|---|---|---|---|
| Russia |  |  |  |  |
| Georgia |  |  |  |  |
| Armenia |  |  |  |  |
| Azerbaijan |  |  |  |  |

## COMPREHENSION AND CRITICAL THINKING

Use the graphic organizer to answer the following questions.

**1. Identify**  In the categories above, where would you list details about how Russia's government is set up? Where would you list details about Armenia's mining industry?

_______________________________________________________________________

_______________________________________________________________________

**2. Interpret**  Of the following seven words (*Azeri, Armenian, Georgian, Baku, Russian, Orthodox, Chechnya*), which two do *not* belong in the category "People/Culture"?

_______________________________________________________________________

_______________________________________________________________________

**3. Sequence**  If you were going to describe the countries by size, from smallest to largest in area, in what order would you describe them?

_______________________________________________________________________

_______________________________________________________________________



     Interactive Reader and Study Guide

# Russia and the Caucasus

**Section 1**

**MAIN IDEAS**

1. The physical features of Russia and the Caucasus include plains, mountains, and rivers.
2. Climate and plant life change from north to south in Russia and vary in the Caucasus.
3. Russia and the Caucasus have a wealth of resources, but many are hard to access.

## Key Terms and Places

**Ural Mountains**  mountain range where Europe and Asia meet

**Caspian Sea**  borders the Caucasus, the world's largest inland sea

**Caucasus Mountains**  mountain range which forms the Caucasus region's northern border with Russia

**Moscow**  Russia's capital

**Siberia**  a vast region in Russia, stretching from the Urals to the Pacific Ocean

**Volga River**  located in western Russia, the longest river in Europe

**taiga**  a forest of mainly evergreen trees covering much of Russia

## Section Summary

### PHYSICAL FEATURES

The continents of Asia and Europe meet in Russia's **Ural Mountains**, forming the large landmass Eurasia. A large part of Eurasia is Russia, the world's largest country in area.

Three much smaller countries, Georgia, Armenia, and Azerbaijan, lie to the south in the Caucasus. This area, which lies between the Black Sea and the **Caspian Sea**, is named for the **Caucasus Mountains**.

Plains and mountains cover much of Russia and the Caucasus. The fertile Northern European Plain, Russia's heartland, extends across western Russia. Here lies **Moscow**, Russia's capital.

Beyond this plain, the vast region of **Siberia** stretches from the Ural Mountains to the Pacific Ocean. The West Siberian Plain is a huge, flat,

**Which two continents meet in the Ural Mountains?**

_______________________

_______________________

**What three countries make up the Caucasus?**

_______________________

_______________________

**What is the capital of Russia?**

_______________________



     Interactive Reader and Study Guide

## Section 1, *continued*

marshy area. East of this plain lies the Central Siberian Plateau. High mountain ranges run through southern and eastern Siberia. Eastern Siberia is called the Russian Far East, which includes the Kamchatka Peninsula and several islands. This area is part of the Ring of Fire, known for its earthquakes and active volcanoes.

**What is the Ring of Fire?**

_______________________________

_______________________________

The Caucasus countries consist mainly of rugged uplands. The Caucasus Mountains contain Europe's highest peak, Mount Elbrus.

Rivers in Russia include the Ob, Yenisey, Lena, and **Volga**, the longest river in Europe. Russia also has some 200,000 lakes, including Lake Baikal, the world's deepest lake. The Black Sea connects to the Mediterranean Sea and is important for trade. The Caspian Sea is the world's largest inland sea.

**What is the longest river in Europe?**

_______________________________

## CLIMATE AND PLANT LIFE

Russia is mainly cold, with short summers and long, snowy winters. Plant life includes small plants in the north and the vast **taiga**, a forest of mainly evergreen trees. Climate in the Caucasus ranges from cooler in the uplands to warm and wet along the Black Sea to mainly hot and dry in Azerbaijan.

## NATURAL RESOURCES

Russia and the Caucasus have a wealth of natural resources, including rich soils, timber, metals, precious gems, and energy resources. These resources have been poorly managed, however, and many remaining resources lie in remote areas.

**What are the region's natural resources?**

_______________________________

_______________________________

## CHALLENGE ACTIVITY

**Critical Thinking: Making Generalizations** Based on what you've read so far, write a short essay about what it might be like to live in Russia.





Interactive Reader and Study Guide

# Russia and the Caucasus

Section 2

**MAIN IDEAS**

1. The Russian empire grew under powerful leaders, but unrest and war led to its end.
2. The Soviet Union emerged as a Communist superpower with rigid government control.
3. Russia's history and diversity have influenced its culture.

## Key Terms and Places

**Kiev** early center of Russia, now the capital of Ukraine

**Cyrillic** a form of the Greek alphabet

**czar** emperor

**Bolsheviks** Communist group that seized power during the Russian Revolution

**gulags** Soviet labor camps

## Section Summary

### THE RUSSIAN EMPIRE

In the AD 800s, Viking traders from Scandinavia helped create the first Russian state of Kievan among the Slavs. Kievan was centered around the city of **Kiev**, now the capital of Ukraine.

Over time, missionaries introduced Orthodox Christianity and **Cyrillic**, a form of the Greek alphabet which Russians still use today.

In the 1200s, Mongol invaders called Tatars conquered Kiev. Local Russian princes ruled several states under the Mongols. Muscovy became the strongest state, with Moscow its main city. After about 200 years Muscovy's prince, Ivan III, seized control from the Mongols. Then in the 1540s, his grandson Ivan IV crowned himself **czar**, or emperor. He became known as Ivan the Terrible for being a cruel and savage ruler.

Muscovy developed into the country of Russia. Peter the Great and then Catherine the Great ruled as czars, building a huge empire and world power.

**What is Cyrillic?**

_______________________________

_______________________________

**What name was Ivan IV given for being such a cruel and savage ruler?**

_______________________________

_______________________________



 Interactive Reader and Study Guide

**Section 2,** *continued*

Unrest, war, and other problems weakened the Russian empire. In 1917, the czar lost support and was forced to give up the throne. The **Bolsheviks,** a Russian Communist group, seized power in the Russian Revolution. Bolshevik leaders formed the Soviet Union in 1922.

> What was the name of the Russian Communist group that seized power during the Russian Revolution?
>
> _______________________
>
> _______________________

## THE SOVIET UNION

The Soviet Union became a Communist country, led by Vladimir Lenin. This type of government controls all aspects of life and owns all property. After Lenin's death in 1924, Joseph Stalin ruled as a brutal dictator. He set up a command economy. His government made all economic decisions, took over all industries and farms, and strictly controlled its people. Anyone who spoke out against the government was sent to **gulags,** harsh Soviet labor camps often located in Siberia.

> Who was the first leader of the Soviet Union when it became a Communist country?
>
> _______________________
>
> _______________________

After suffering many losses in World War II, Stalin worked to create a protective buffer around the Soviet Union. He set up Communist governments in Eastern Europe. As a result, the Cold War, a period of tense rivalry and arms race between the United States and the Soviet Union, developed. The Soviet Union grew weak in the 1980s, then it fell apart in 1991.

> What was the Cold War?
>
> _______________________
>
> _______________________

## CULTURE

More than 140 million people live in Russia. Most are ethnic Russians, or Slavs, but Russia also has many other ethnic groups. The main faith is Russian Orthodox Christian. Russia has made many contributions to the arts and sciences, including ballet and space research.

> How many people live in Russia today?
>
> _______________________
>
> _______________________

## CHALLENGE ACTIVITY

**Critical Thinking: Drawing Inferences** What do you think Russia would be like today if the Soviet Union had not collapsed? Give support for your answer.

# Russia and the Caucasus

### Section 3

**MAIN IDEAS**

1. The Russian Federation is working to develop democracy and a market economy.
2. Russia's physical geography, cities, and economy define its many culture regions.
3. Russia faces a number of serious challenges.

## Key Terms and Places

**dachas**  Russian country houses

**St. Petersburg**  city founded by Peter the Great and styled after those of Western Europe

**smelters**  factories that produce metal ores

**Trans-Siberian Railroad**  the longest single rail line in the world, running from Moscow to Vladivostok on the east coast

**Chechnya**  a Russian republic in the Caucasus Mountains, an area of ethnic conflict

## Section Summary

### THE RUSSIAN FEDERATION

The Russian Federation is governed by an elected president, an appointed prime minister, and a legislature called the Federal Assembly. Since 1991, Russia has been changing from communism to a democracy with a market economy based on free trade and competition. These changes have led to economic growth that is most visible today in Russia's cities. A wide range of goods and services can be found here. Most Russians live in large apartment buildings in cities, but they still enjoy nature. Cities often have large parks and many richer Russians own **dachas**, or country houses.

> **What is the Russian legislature called?**
> ___________________

### CULTURE REGIONS

The Russian heartland consists of four major culture regions—the Moscow, St. Petersburg, Volga, and Urals regions.

> **What are the four regions of Russia's heartland?**
> ___________________

### Section 3, *continued*

Moscow is Russia's capital and largest city. The Kremlin holds Russia's government offices, beautiful palaces, and gold-domed churches. Moscow is a huge industrial region.

Founded by Peter the Great, **St. Petersburg** was styled after cities of Western Europe and served as Russia's capital for some 200 years. Its location on the Gulf of Finland has made the city an important trade center.

> **Where is St. Petersburg located?**
> ____________________________

The Volga River of the Volga region is a major shipping route and source of hydroelectric power. Factories here process oil and gas. The Caspian Sea provides black caviar.

> **Which river is a major shipping route in the Volga region?**
> ____________________________

The Urals region is an important mining region. **Smelters**, or factories that process metal ores, process copper and iron.

Siberia is east of the Urals. Winters there are long and severe. Siberia has many natural resources, but accessing them is difficult. Lumber, mining, and oil production are the most important industries. Most towns follow the route of the **Trans-Siberian Railroad**, the longest single rail line in the world.

> **What are winters like in Siberia?**
> ____________________________
> ____________________________

> **What is notable about the Trans-Siberian Railroad?**
> ____________________________
> ____________________________

Siberia's coastal areas and islands are known as the Russian Far East. Resources include timber, rich soils, oil, minerals, and fishing. Vladivostock is the area's main seaport.

## RUSSIA'S CHALLENGES

Russia faces economic, employment, health, population, environment, and ethnic challenges. A major ethnic conflict is in the Russian republic of **Chechnya**, where some people desire independence.

## CHALLENGE ACTIVITY

**Critical Thinking: Drawing Inferences** Based on what you've learned about each region, which area do you think holds the most potential economically for Russia? Why?

# Russia and the Caucasus

**Section 4**

> **MAIN IDEAS**
> 1. Many groups have ruled and influenced the Caucasus during its long history.
> 2. Today the Caucasus republics are working to improve their economies but struggle with ethnic unrest and conflict.

## Key Places

**Tbilisi**  the capital of Georgia

**Yerevan**  the capital of Armenia

**Baku**  the capital of Azerbaijan, center of a large oil-refining industry

## Section Summary

### HISTORY

The Caucasus lies in the Caucasus Mountains between the Black and Caspian seas. The region reflects a range of cultural influences and at one time or another has been ruled or invaded by Persians, Greeks, Romans, Arabs, Turks, Mongols, and Russians.

**Which groups have controlled the Caucasus?**

_______________________

_______________________

In the early 1800s, Russia took over much of the Caucasus, but the Ottoman Turks held western Armenia. Before and during World War I, hundreds of thousands of Armenians were killed by the Turks. After the war, Armenia, Azerbaijan, and Georgia were independent, but became part of the Soviet Union in 1922. They regained independence when the Soviet Union fell in 1991.

### THE CAUCASUS TODAY

Although the region has a long history, the Caucasus countries have had to create new governments and economies. Progress has been slowed by ethnic unrest and conflicts. Each country's government has an elected president, an appointed prime minister, and an elected parliament, or legislature.



     Interactive Reader and Study Guide

**Section 4, *continued***

Georgia is located between the Caucasus Mountains and the Black Sea. **Tbilisi** is the capital. About 70 percent of the people are ethnic Georgians and belong to the Georgian Orthodox Church. Georgian is the official language.

Georgia has struggled with unrest and civil war. Georgians forced out their president in 2003.

Georgia's economy is based on services and farming. Other industries include steel, mining, wine, and tourism.

> **What are Georgia's main industries?**
> ________________________
> ________________________

Armenia is a small, landlocked country south of Georgia. **Yerevan** is the capital. Most of the people are Armenian and belong to the Armenian Orthodox Church.

Armenia fought a war with Azerbaijan in the early 1990s. The war involved an ethnic Armenian area of Azerbaijan that is still controlled by Armenian forces today. This conflict has hurt Armenia's economy, but international aid is helping.

> **What country was Armenia at war with in the early 1990s?**
> ________________________

Azerbaijan is east of Armenia. The Azeri make up 90 percent of the population and are mostly Muslim. Oil is the most important part of the economy. **Baku**, the capital, is the center of this industry. Problems include corruption, poverty, and refugees as a result of the conflict with Armenia.

> **What is the most important part of Azerbaijan's economy?**
> ________________________

## CHALLENGE ACTIVITY

**Critical Thinking: Compare and Contrast** Describe the similarities and differences between the Caucasus countries.



  Interactive Reader and Study Guide